SUCCESSFUL ORGANIC PEST CONTROL

SUCCESSFUL

ORGANIC PEST CONTROL

Environment-friendly ways to deal with unwanted garden pests and diseases

Trevor G. Forsythe

Thorsons Publishing Group

First published 1990

British Library Cataloguing in Publication Data

Forsythe, Trevor G.
Successful organic pest control.
1. Gardens. Pests. Biological control
I. Title
635.04996

ISBN 0-7225-2242-8

Published by Thorsons Publishers Limited, Wellingborough, Northamptonshire NN8 2RQ, England

Typeset by Harper Phototypesetters Limited, Northampton, England
Printed in Great Britain by Mackays, Chatham, Kent

1 3 5 7 9 10 8 6 4 2

Dedication

John R. McInnes
1947-1984

About the author

Dr Trevor G. Forsythe is a professional naturalist, author and research entomologist. He has contributed to numerous scientific journals, featured in BBC *Wildlife Magazine*, contributed to BBC *Wildlife* programmes and is engaged in research into plant molluscides, for the control of slugs, and feeding in ground beetles and other predators. His first book, *Common Ground Beetles*, is one of a very successful and respected series of *Naturalists Handbooks* (published by Richmond Publishing Co. Ltd.). He is an honorary research fellow of both Manchester University where he graduated, and Lanchester Polytechnic in Coventry.

Dr Forsythe is an Assistant Keeper in Biology at the Leicestershire Museum, Art Gallery and Records Service where he helps curate the insect and other collections.

Contents

Acknowledgements

I would like to thank the authorities of the I.H.R., Wellesbourne for use of library facilities; David Woodruff and Bob Ellis; Dr R. Cooper, Department of Biological Sciences, Coventry (Lanchester) Polytechnic for use of laboratory facilities; the National Centre for Organic Gardening, Ryton-on-Dunsmore; my colleagues in the Biology section at the Leicestershire Museum, Arts and Records Centre, Leicester for their continued support; and Dr S. E. R. Bailey, Dr M. E. G. Evans and Gordon Blower, Department of Environmental Biology, Manchester University.

1
Friend or foe?

At first glance the common wasp (*Vespula vulgaris*) appears to have little to commend it. Wasps either congregate around litter bins or plague us when we are trying to enjoy a quiet cup of tea in the garden. They are very much maligned creatures, probably because we do not appreciate being stung. Left alone to continue their frantic quest for food, they do us very little harm and will not attack unless provoked — running around waving your arms in the air and screaming is probably the worst reaction if you want to avoid being stung. Only in September when the soft fruit season is in full swing do wasps tend to make themselves a nuisance.

They are perhaps one of the most beneficial insects in the garden, because they rear their offspring mainly on insect larvae — many of which are harmful. When you consider that a wasp nest will house anything up to 30,000 wasps — each one reared mainly on insect larvae — that could represent a considerable saving in leaf area to the jaws of caterpillars.

But it is not only the social wasps which are beneficial.

Fig. 1 Common wasp

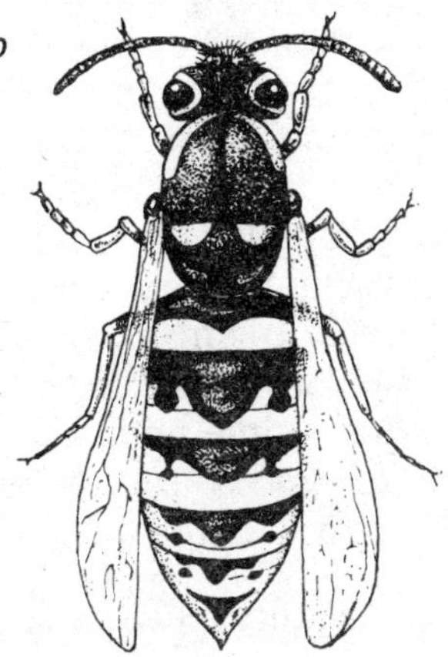

Solitary wasps, although much smaller and fewer in number, live a solitary life, as their name suggests, provisioning their nests with prey as food for their offspring. Many of these solitary wasps specialize in catching certain types of prey. Some species prey on aphids and some on weevils, whilst others hunt caterpillars.

Convicing people that the wasp may be beneficial can be an uphill task; to convince them that some species of aphid may also be beneficial is more or less impossible. Aphids are on everybody's blacklist of dreaded pests and gardeners, including myself, often wage war against them almost at any cost. It would be stretching the credulity of most people to ask them to accept the aphid as the gardener's friend! Nevertheless, the net impact of populations of certain aphids on crops may be beneficial.

It seems that the honeydew, which is mainly sugar and which continually rains down onto the ground from aphid infected plants encourages the number of Azotobacter bacteria living in the soil to multiply, and that these in turn seem to convert atmospheric nitrogen into a form available to plants. It is thought that the gain in available nitrogen exceeds the loss in sugar and is therefore of net benefit to the plant. (Of course, the aphid colony can grow to be so large that it becomes counter-productive.)

Ground beetles are familiar to the gardener and very few people would argue with the fact that they are the gardener's friend. Most of them are sombrely coloured, and most (although not all) are nocturnal. They spend the daytime hiding under stones or other garden debris. The gardener is most likely to come across them whilst digging. They are easily recognized as black, rapid-running beetles with long, slender legs and powerful, sharply pointed jaws.

Most species are omnivorous which means they eat food of both animal and plant origin. Quite a few are strictly carnivorous but some consume mainly plant material. What is often not realized is that the majority of ground beetle larvae are predominantly predatory and may even be more beneficial than the adults. Most adults are opportunists — in other words, they will attack and consume any available prey. The kind of prey eaten is almost entirely regulated by the size of the prey and the size of the beetle's jaws. As most ground beetles range in size from a few millimetres to over 30 mm (1¼ in.) they present a formidable array of predators capable of tackling most kinds of arthropod pest.

However, there are a few destructive beetles to watch out for, although they represent fewer than 1 per cent of all ground beetles. The strawberry seed beetle (*Harpalus rufipes*) and strawberry ground beetles (*Pterostichus madidus* and *Pterostichus melanarius*) are known to the gardener for the damage they do to strawberries. The preferred food of the strawberry seed beetle are seeds and plant material of fat hen, chickweed and other weed plants, while the strawberry ground beetles include slugs, and other arthropods in their diet.

The extent of damage to strawberries by *Pterostichus madidus* appears to be determined not only by hunger and thirst, but also by factors which affect the beetle's activity. The beetle is more active after rain, and inactive during periods of drought. A dry period will therefore result in the beetle becoming both starved and dessicated. When it starts raining again it will feed readily, so that attack on the fruit is likely to be most severe at this time.

As these beetles are also three of the most common and widely ranging species of ground beetle, it may be wise to consider them as 'beneficial' and only try to control them actively where strawberries are being cultivated.

The destruction of hedgerows and associated vegetation by their removal and use of weedkillers over the last few decades is thought to have greatly reduced ground beetle numbers. As

Fig. 2 Strawberry seed beetle

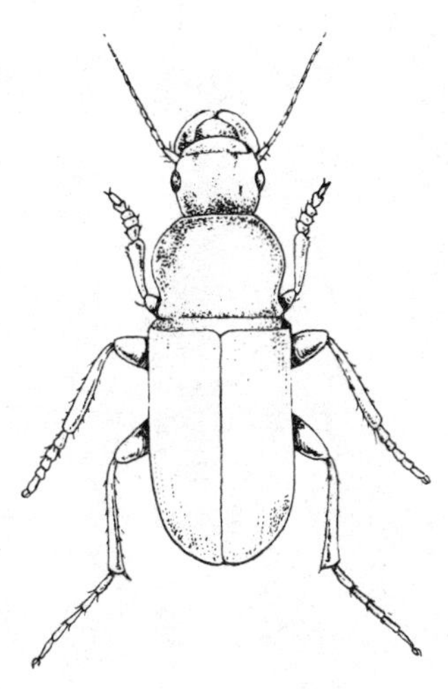

ground dwellers these beetles and their larvae are exposed for long periods to residues left by pesticides; they are also mainly predators and the effect of eating contaminated prey may be cumulative.

Earwigs are considered as pests by most gardeners and commercial growers of fruit. In orchards earwigs feed most readily on the fruit where the skin has already been broken. Because of these habits they often introduce fungal infections when they feed. Therefore even small amounts of damage may result in losses of fruit, particularly when in store, through the spread of fungal diseases.

Most gardeners, however, notice very little damage as a result of earwigs unless they grow chrysanthemums or dahlias. The choice of food for the earwigs is closely related to the fact that they are tactile animals and need to feel hemmed in by their surroundings. Chrysanthemum and dahlia flowers satisfy this need; during the day the earwigs hide in the heart of the bloom and at night appear to feed. As they are omnivorous insects, they will eat the first palatable thing that comes to hand — your dahlias or chrysanthemums. However, what is often not realized is that they also readily eat green algae growing on the branches of trees, moss, fungi and pollen; in addition they prey on other insects, including aphids and codling moth eggs. It seems that earwigs may act as natural biological control agents, after all.

If you are not yet convinced of the benefit of having earwigs around, then you may be interested to hear that they can be

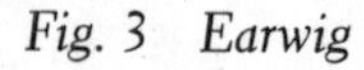

Fig. 3 Earwig

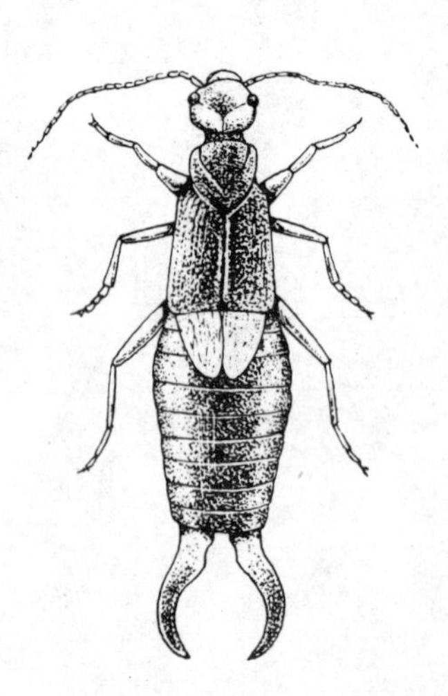

loving parents. The pair hollow out a brood chamber a few inches below the soil or under a stone or piece of wood where a small clutch of eggs is laid. Then the male usually departs and the female stays with the eggs, spending her time licking and turning them, probably in order to prevent fungal spores germinating and growing on the developing eggs. If the eggs are disturbed she will gather them together and move them to a new site. After they hatch, she continues to provide maternal care, repelling predators and feeding her young.

It is not only insects that have suffered from a bad press; other arthropods and animals have suffered, too. Woodlice are usually blamed for devouring young leaves of a wide range of flowering plants during the night — particularly in the greenhouse or conservatory. In fact, woodlice prefer rotting plants or animal material. Consequently they recycle nutrients and perform a very useful function by speeding up the decaying process and increasing soil fertility. Unfortunately, woodlice do have a penchant for seedlings, eating holes in the leaves and gnawing at the stems.

The snake-like appearance of the slow-worm has often led to its persecution. The slow-worm is in fact not a snake but a legless lizard. It is neither slow as its name suggests nor poisonous. Its favourite prey are slugs, but it will also eat worms, soft-bodied insects, larvae and spiders.

Slow-worms are rarely seen in the daytime except when basking in the sun, but prefer to hunt at night when their prey are most active. They will reach up to take slugs feeding on plants and will burrow down into the soil to capture slugs that live underground.

Hedgehogs, on the other hand, are considered the gardener's friend, probably because they eat the dreaded slug, but then doesn't the slow-worm? Hedgehogs, like all creatures, have a chink in their armour — they like the occasional fledgling, so lock up your baby chicks at night!

Birds can be both friend and foe to the gardener, in that some species cause most damage during the winter months when there is little for them to eat. They will attack buds of fruit, particularly blackcurrants, cabbages and other brassicas. However, during the summer months birds also control grubs, caterpillars, slugs, aphids and other pests.

The underlying premise for maintaining a good organic environment is *diversity*. The greater variety of both plant and animal life the more sustainable it is likely to be. The animals and plants we often classify as 'pests' or 'weeds' are part and parcel of a dynamic self-regulating system with the surrounding environment. Pests cannot be eliminated entirely. Controlling a pest with a predator has one major disadvantage: once the pest is eradicated, the predator also dies out because it has nothing left to eat. Even pests, then, are essential to the scheme of things. Nature does not eradicate a plant or animal but maintains, instead, a kind of balance between the opposing sides. Only when nature becomes unbalanced do the scales shift and a new equilibrium is established.

Gardening with nature is one of understanding the environment, working with it rather than battling against it, accepting pests and weeds as part of a self-regulating system and only intervening when populations of pests and weeds get out of hand. Try to look at your garden as a living entity and not as a group of plants and animals living in isolation.

2
Garden friends and how to encourage them

It is not always possible to divide insects and other creepy-crawlies into those that harm plants and those that are useful in the garden. The majority of insects are neither harmful nor beneficial. Some are both — wasps are pests when they annoy us or eat ripening fruit, but not when they eat harmful insects. Similarly, weeds are only weeds when they are growing in places where we do not want them to grow.

Insects pass through two main stages in their life cycle when they may be harmful: an immature stage either as a caterpillar, grub, maggot, larva or nymph and an adult, often winged, stage. There is usually an inactive stage in-between — the pupal stage as between caterpillars and butterflies, and an egg stage. An insect may only become a pest at one stage during its life cycle. The adult crane-fly or daddy-long-legs, for instance, does no harm, but its larva the leather-jacket feeds on the roots of turf and bedding plants. The same goes for the click beetle, which does little harm as an adult. The larvae, however, are the wireworms that attack the roots of vegetables and other plants.

A lot of insects, other arthropods and small creatures are beneficial most of the time. Many garden plants, for instance, would not set seed or produce fruit without the help of pollinating insects and other animals. Apart from hive bees, many kinds of flies, bumble-bees, wasps and beetles help pollinate flowers. The bright yellow, oil-seed rape would not produce its crop of seed without pollinating insects. It is important, therefore, to confine the use of insecticides to the evening when bees are less active and not to spray plants that are in flower. This is not always possible, of course, because the target insect, the pest, may only be present during daylight hours. Insecticides derived from plants are also contact insecticides and therefore must be directed at, and must come into contact with, the pest to be effective.

Many garden creepy-crawlies have an important role to play in tidying up our gardens. Several types of beetle specialize in burying dung or dead animals and birds. They are often aptly named as dung beetles or burying beetles. The maggots of the bluebottle consume the corpses of many dead birds and animals. Numerous organisms such as bacteria and fungi live on dead or decaying plant material. All these organisms, therefore, play a vital role in returning nutrients to the soil.

A garden without birds, animals or insects is a very unattractive place. It is the drone of bees, the chirping of grasshoppers and the songs of birds that bring the garden alive and add that extra dimension. Butterflies and dragonflies not only add colour, but also bring movement to the garden.

If caterpillars eat your plants, pick them off or leave them to their natural enemies — the criterion is to intervene with the spray-can as a last resort rather than adopt the philosophy so prevalent in the 1960s of 'If it moves spray it'. Many garden inhabitants like hedgehogs, frogs, toads and certain birds live almost entirely on insects and other garden creepy-crawlies — they will not come at all if there is nothing for them to eat.

Most garden pesticides kill a wide range of bugs and indiscriminate use of pesticides will kill the 'good bugs' too.

Beetles

Ground beetles

There are approximately 350 species of ground beetle in the British Isles. All are beneficial insects to varying degrees, apart from three species which damage strawberries (see page 76). Even the strawberry pests are only a nuisance when their numbers become unusually large and then only in strawberry beds.

Several species of ground beetle are omnivorous, some are carnivorous and a few are herbivorous. The larvae may be even more important than the adults because they are thought to be mainly predators.

Several small (4-5 mm) species of *Bembidion* and *Trechus*, for instance, are important predators of the immature stages of the cabbage root fly. Because they occur in quite large numbers and are very active predators, they can have a considerable impact on the numbers of cabbage root fly eggs.

The type of protective disc used to control cabbage root fly (see page 68) is also attractive to predatory beetles, because of the humid microhabitat created underneath the disc.

Some of the larger (20-30 mm/¾-1¼ in.) ground beetles — the violate ground beetle, for instance — are major predators of small slugs, such as the field slug. Perhaps one of the best ways of protecting vulnerable plants from the ravages of slugs is to plant them in what is in reality a giant pitfall trap (see page 58). Beetles fall into the trap and cannot get out and therefore eat the slugs feeding on the plants.

Most ground beetles are opportunists in that they will feed on any anthropod which comes their way; the only limiting factor governing the type of prey taken is perhaps size. They will feed on aphids, small slugs and snails, insect eggs, small flies, springtails, mites, ants, nematodes and most other creepy-crawlies.

Perhaps the best way to encourage ground beetles is by not using synthetic pesticides such as aldrin, which have a devastating effect on beetle populations. Ground beetles are top of the food chain in the creepy-crawly world and therefore accumulate pesticides. The resulting reduction in their numbers produces an inevitable increase in the pest.

Roof slates or similar light slabs left around the garden provide shelter for the beetles during the daytime, whilst logs and tree stumps provide overwintering sites. Ground cover, such as that used for intercropping, encourages ground beetles.

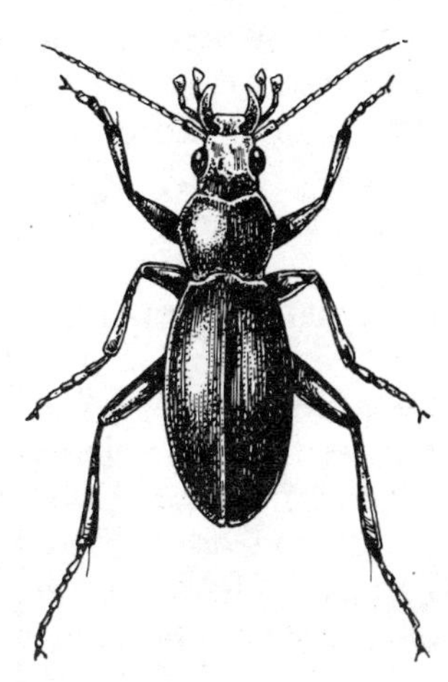

Fig. 4 Ground beetle — violate ground beetle

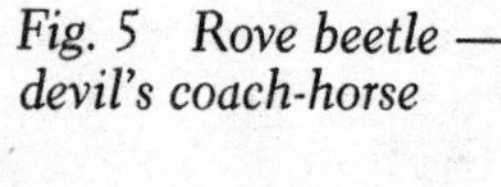
Fig. 5 Rove beetle — devil's coach-horse

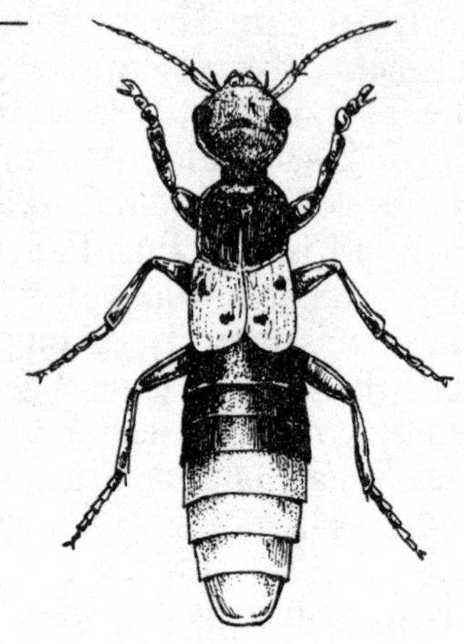

Rove beetles

The rove beetles comprise another large family of over 950 species of beetles. Whereas the ground beetles are mainly surface hunters, the rove beetles hunt in cracks in the soil or in the litter layer. They are beautifully adapted for this way of life with their powerful jaws and highly flexible bodies. The best known example to the gardener is the devil's coach-horse.

Like the ground beetles, rove beetles are a major predator of cabbage root fly eggs. They are also important predators of lettuce root aphid, strawberry aphid, red spider mite and numerous other pest species. Their larvae are also often predatory.

As with ground beetles, rove beetles are susceptible to synthetic pesticides which should not be used. Similar methods to those used to encourage ground beetles should be adopted in order to increase or sustain populations of rove beetles.

Ladybird beetles

There are approximately 40 species of ladybird in the British Isles. All species are predatory both as adults and larvae, except one species which feeds on leguminous plants and can be a minor pest occasionally. When alarmed, ladybirds emit a bitter fluid from the leg joints (reflex bleeding) which appears to be distasteful to birds and other predators. Their conspicuous appearance is an example of warning coloration.

They prey on a variety of pests such as aphids, scale insects, mealy-bugs, thrips, small caterpillars and mites, although a few species are specific in their choice of food. Both the adults and larvae eat many hundreds of aphids and other pest species during their lifetime.

Ladybirds pass the winter as adults, sheltering under loose bark, in crevices in walls and trees, in hollow plant stems and anywhere sheltered and more or less dry. This knowledge of overwintering sites can be put to good use by the gardener. Beetles will hibernate in hollow canes if they are set in sheltered spots. Alternatively, certain 'weeds' such as dock or umbellifers can be grown and dead-headed before they set seed in order to produce the hollow stems for hibernating beetles. I have used this to great effect in my own garden.

At hibernation time many species are gregarious, collecting in their winter quarters, sometimes in vast numbers. The same winter quarters are often used year after year and the beetles are probably attracted by the characteristic odour of their own kind. In the spring they leave their hibernation sites and become solitary insects once more. These sites should be preserved and provided in order to build up ladybird populations.

There is some evidence to suggest that vegetation height plays a part in habitat selection by ladybirds. It may be worth while noticing which wild habitat conditions most favour their successful breeding. If similar conditions are created in the garden, then ladybirds will be encouraged to breed, thus producing adequate supplies of young which in turn will be

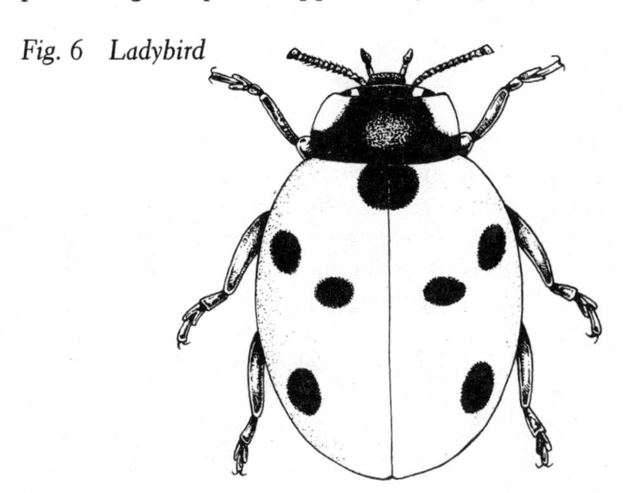

Fig. 6 Ladybird

available to migrate to crops. For instance, ladybirds leaving hibernation often find their way to nettles infected with aphids where they produce the 'nettle brood' from which newly emerged adults disperse and colonize other plants in the vicinity.

When spraying against aphids, care must be taken not to use derris if ladybirds are present as it is highly toxic to beneficial insects, especially ladybird larvae and eggs. Pyrethrum is also toxic to ladybirds.

Perhaps the best and least toxic spray to beneficial insects is insecticidal soap. Nicotine, however, is very effective against aphids, but while not toxic to ladybirds it has the disadvantage of being highly toxic to bees, so should only be used in the evening. (Nicotine is now available only to professional growers.)

Other beetle friends

There are many other predatory beetles too numerous to mention here, but the glow-worm is perhaps the best known and most likely to be observed by the gardener.

The glow-worm is in fact the larval stage of a beetle that feeds on tiny snails and small slugs. A conspicuous spot of light announces its presence on grassy banks, in fields or hedges at dusk during the months of June and July.

The glow-worm lays its eggs on the ground in the vicinity of snails and, as soon as they hatch, the larvae make their way into the snail shell in order to devour the inhabitants — a habit which is also paralleled by the adult beetle.

Hoverflies

Hoverfly larvae can be divided into four categories regarding feeding: species that feed on plant tissues and sap; species that scavenge decaying matter; species that live in nests of social insects such as wasps and ants, and carnivorous species where the normal food is aphids. It is the carnivorous species which are of main interest to the gardener. They nearly all belong to one subfamily, Syrphidae.

The aphid-eating larvae can themselves be divided into two groups: obligate aphid-feeders that can only develop normally by feeding on aphids or other suitable prey, and faculative aphid feeders which, although they prefer aphids, can complete their development by including some rotting plant material in their diet.

Most obligate carnivorous species have catholic tastes in aphids — most common species fall into this category. Some species, however, avoid certain aphid species.

Each larva can eat up to 1200 aphids during the course of its development, obligate aphid-feeders being in general more efficient predators than the faculative species.

Hoverfly larvae frequently decimate colonies of aphids, resulting in periods of starvation. Unusually small adults are then produced and in consequence the number of eggs laid by the female is reduced. This in turn leads to an increase in aphid populations and the cycle starts again. This is all part of the natural balance that occurs in nature.

Adults feed on pollen and nectar. Females, and probably males also, emerge with undeveloped reproductive systems and must feed to bring them to maturity. The proteins and amino acids of pollen appear to be essential for maturity. Nectar contains only trace amounts of these substances and probably does not play an important role in the development of the reproductive system. It is the females, however, which require large amounts of proteins and amino acids if they are to produce successful batches of eggs.

Nectar is almost a pure solution of sugars, and is thus required for use as energy during flight.

The mouthparts of different species are adapted to feed on different types of flowers. Short broad mouthparts are adapted

Fig. 7 Hoverfly

for pollen feeding whereas long, thin mouthparts are adapted in order to reach the nectar in flowers with long corollae.

Hoverflies appear to select flowers predominantly by their colour; scent and other factors play only a minor role. Most hoverflies are recorded as selecting plants with yellow or white flowers. Attraction to these colours, however, does not preclude flowers of other colours; hoverflies have been recorded visiting creeping thistle, knapweed and vetches as well as plants which produce yellow or white flowers such as broom, umbellifers and flowers of the daisy and rose family.

By planting species such as buckwheat (*Fagopyrum esculentum*), yarrow (*Achilles millefolium*), candytuft (*Iberis umbellata*), French marigold (*Tagetes* spp.) and native wild flowers with white or yellow flowers, hoverflies can be attracted to the garden where they will search out aphid colonies in which to lay their eggs.

Egg-laying behaviour has been studied most extensively in species with aphid-feeding larvae. Females of different species respond to various stimuli such as the size and density of the patch of plants, the colour, form and smell of the flowers, the smell of aphids, of honeydew, the size and position of the colony of aphids and green colours which lead the female to young plant growth where aphids are more likely to occur.

Both derris and nicotine are reputed to be harmless to hoverfly larvae, but highly toxic to aphids. However, care must always be taken when spraying in the presence of the adult hoverflies and at the first signs of prolonged damage to hoverfly larvae populations, spraying should cease.

Other predatory flies

These include robber or assassin flies which hunt grasshoppers, beetles, bees, wasps, ichneumons and flies; gall midges, one species of which preys upon the fruit-tree red spider mite; species that prey on aphids; and flies of the family Sciomyzidae which include a number of species whose larvae prey upon snails and slugs (see page 57).

Parasitic flies

Tachinid flies are 'bristly' parasitic flies which include the blowflies and greenbottles as well as many lesser known flies.

Their larvae are either internal parasites of invertebrates

such as snails, earthworms, grasshoppers, beetles, bugs, caterpillars of butterflies or moths and woodlice; or external parasites of mammals or birds and include familiar insects such as greenbottles or sheep blowflies.

Lacewings

There are several families of lacewings, but the green lacewings, brown lacewings and powdery lacewings are the most beneficial to the gardener.

The green lacewings, known also as golden eyes, are useful in the control of greenfly. Green lacewings lay their eggs scattered over plants without regard to the presence of aphid colonies. The larvae, however, are active predators and with the aid of their long jaws seek out small aphids hidden in the folds of young leaves, often where ladybirds fail to reach them. Their curved jaws are like hollow calipers with which they seize aphids and suck out their body fluids. Adult green lacewings lay many hundreds of eggs and each larva eats many hundreds of aphids during its development. Most species are not obligate aphid-feeders but will also prey on mites, small spiders and leaf-hoppers.

The brown lacewings are generally smaller than the green lacewings and both adults and their larvae also feed on aphids, thrips, mites and the like.

Powdery lacewings, as their name suggests, are covered with a white powdery substance and somewhat resemble glasshouse whitefly. The larvae prey upon small aphids, scale insects and mites.

Fig. 8 Lacewing

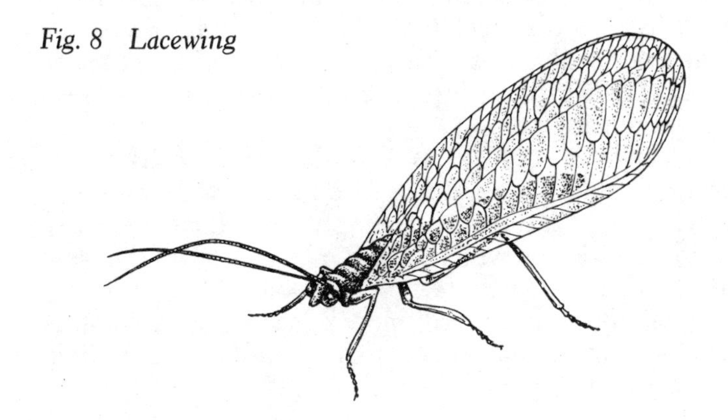

Certain species are predators of the fruit-tree red spider mite. In orchards the larvae and adults feed almost entirely on mites and their eggs. The larva pierces fruit-tree red spider mites with its jaws and sucks out the body contents.

The larvae may be active even during the autumn, when they live almost entirely on the winter eggs of the mites, reducing the numbers of eggs quite considerably.

Before 1923 the fruit-tree red spider mite was unknown as a pest. There can be little doubt that the increase in numbers of this pest was due to the adoption in the 1920s and more recently of new and more powerful insecticides. The fruit-tree red spider mite is subject to predation by numerous species of capsid and other bugs, lacewings, ladybirds and predatory mites. Winter washes of tar oil often destroy these predators, enabling spider mites to increase in number.

Derris is toxic to adult green lacewings, although it has a low toxicity to the larvae; nicotine, however, is reputed to be harmless to both adults and larvae of the green lacewing, but is now available only to professional growers.

The adult green lacewings feed also on pollen, nectar and honeydew and can be attracted to the garden by an artificial diet of wheast, which is made up of yeast, sugar and water. This is attractive to beneficial insects such as lacewings and by placing wheast around the garden you will encourage adult lacewings to stay in the area and to lay eggs.

Bugs

Bugs (Hemiptera) do not have a pupal (chrysalid) stage; the young resemble wingless adults and are generally called nymphs. They have piercing mouthparts which they use to puncture tissues to extract food. Many, such as aphids, are important pests of crops; others, such as the black-kneed capsid, are predatory on insects and mites and are a great benefit to the gardener.

The black-kneed capsid is probably one of the most important predatory bugs. It lays its eggs on a wide variety of trees, including fruit trees. The nymphal period lasts between 25 and 35 days, during which time the bugs prey on the fruit-tree red spider mite. In fact both adult and nymph are predatory on the fruit-tree red spider mite, but will also attack various other mites, small insects such as leaf-hoppers, aphids, thrips and caterpillars.

It has been shown that the black-kneed capsid makes a marked impression on infestations of fruit-tree red spider mite provided no insecticides are used. Anthocorid bugs also occur in orchards and prey on aphids, scale insects, apple suckers, caterpillars, apple-blossom weevils and mites.

Anthocorid bugs are also important predators of lettuce-root aphid. This aphid feeds on the roots of lettuce. It overwinters as eggs on poplar and the young live in galls on the leaf stalks. In July the galls split open to release winged aphids which disperse to lettuce. Anthocorid bugs will enter the open galls and kill the entire population before they disperse. It seems that the adult bugs lay their eggs outside the gall in time for the young to penetrate the opening galls.

Many other bugs are predatory, including shield bugs, nabid bugs and reduvid bugs.

Tar oil used as a winter wash on fruit trees is particularly toxic to overwintering bugs such as *Anthocoris nemorum*. However, there appears to be some dispute as to whether tar oil is toxic to the eggs of the black-kneed capsids.

Nicotine, which can be used against aphids and the apple sawfly in the form of nicotine sulphate, has little effect on predatory capsid and anthocorid bugs, likewise ryania. Derris, on the other hand, kills the nymphs of the black-kneed capsid.

Predatory mites

The nymphs and adults of predatory mites attack and feed on other mites, such as spider mites, apple and plum rust mites and the like. They can often survive by feeding on a diet of pollen grains, fungal spores and plant vegetation. The variety of food eaten by these predatory mites means that they are not dependent on any particular prey so, although they do not become more numerous as their prey increases, they will remain on the tree when their prey is in short supply, unlike insects which tend to move off elsewhere when they have decreased their prey to low numbers.

Predatory mites are most important on fruit trees and in biological control in greenhouses (see Chapter 4).

Both nicotine (available only to professional growers) and ryania are harmless to several species of predatory mites, although some species are susceptible to derris as they are to the fungicide dispersible sulphur.

Wasps

Parasitic wasps

Most parasitic species lay their eggs on or inside the larvae, pupae or eggs of the host by means of a long, needle-like ovipositor which rises from the underside of the abdomen.

The ichneumon flies are perhaps the best known of the parasitic wasps. For the most part they parasitize caterpillars of butterflies and moths, often quite heavily. For instance, 90 per cent of the caterpillars of the diamond-back moth can be parasitized. Some species parasitize sawflies and aphids whereas others parasitize spiders, beetles and lacewings. Most species are internal parasites although some are external parasites, laying their eggs on the outside of the host.

The Braconids also parasitize a wide range of insects, especially butterfly and moth caterpillars, but also more importantly aphids. *Apanteles glomeratus* is a common parasite of the caterpillars of the large white butterfly (*Pieris brassicae*). Clusters of sulphur-yellow cocoons are found on and around the collapsed skin of the host. One caterpillar may contain as many as 150 parasitic larvae.

One group of Braconids attacks aphids. The empty aphid skins remain on the plant long after the parasites have left the host and are an indication of their presence. There are Braconid parasites of the cabbage aphid, peach-potato aphid, potato aphid and many other aphids.

Braconid wasps also parasitize both the larvae and pupae of the carrot fly and have two generations a year, coinciding with those of the host.

Parasitic wasps are less likely to be driven off by ants in attendance on aphids, unlike many other natural enemies such as ladybirds, hoverflies and lacewing larvae.

The Chalcids are probably more important than either the ichneumons or the Braconids as natural enemies of insect pests. *Encarsia formosa* is important economically in the control of whitefly in greenhouses (see Chapter 4) and is probably the best known of the Chalcids.

Chalcids are among some of the smallest known insects, the majority of which are either parasites or hyperparasites (a parasite which lives on or in another parasite). They most commonly attack the larvae, eggs and pupae of butterflies and moths, flies, scale insects and other bugs.

One of the commonest Chalcids parasitize the white butterflies, while others attack scale insects and aphids, hoverflies and blowflies and the strawberry tortrix moth.

Cynipid wasps

One of the most important cynipid wasps is *Idiomorpha rapae*, a parasite of the cabbage root fly. It lays its eggs on the first or second stage larvae; the adult emerging from the puparium (same stage and similar to a pupa, showing lines of segmentation but without indication of legs or wings as is found in moth or beetle pupae).

Solitary wasps

We are all familiar with the black and yellow wasps; social insects where individuals co-operate with their own kind in order to maintain a colony of hundreds or thousands. The solitary wasps, on the other hand, do not collaborate in this way. Each individual works alone. In most species an adult wasp — usually the female — prepares a nest, provisions it with paralysed prey as food for her offspring and deposits an egg, after which the nest is sealed and the adult wasp leaves to provision another nest.

Depending on species, the nest is furnished with different kinds of prey — aphids, weevils, flies, solitary bees, caterpillars, spiders and the like.

Social wasps

A wasp nest may contain as many as 30,000 individuals, all of which will have been reared on insects — many of them harmful. They are truly beneficial creatures, so it is not a good idea to kill them, especially the queen which is usually observed in the spring.

Although they are carnivorous, adult wasps like sweet things, but their short tongues do not allow them to reach the nectar from flowers — hence their attraction to fruit and other sweet things.

Wasps are often considered a nuisance but can be manipulated to act as bio-control agents. Their reliance on lepidopterous larvae as a food source can be utilized by encouraging them to build nests especially near brassica plots where they will prey on large numbers of cabbage butterfly caterpillars.

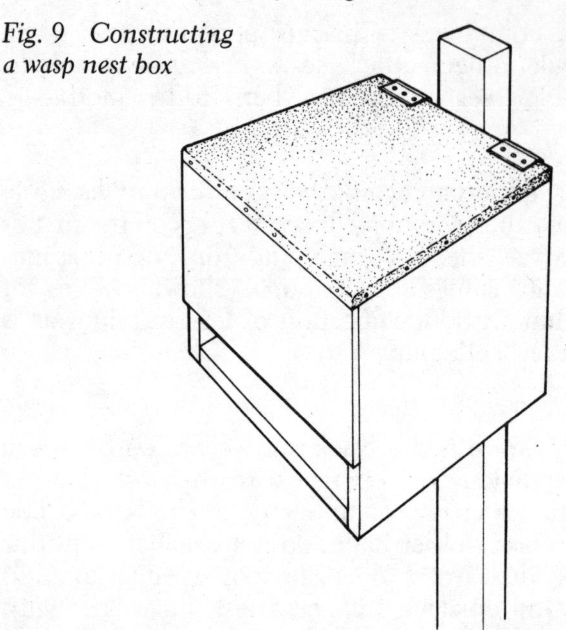

Fig. 9 Constructing a wasp nest box

Nest boxes can be constructed of 2 cm (¾ in.) pine — untreated with preservative or paint — with inside dimensions large enough to house a small nest (35 cm × 35 cm × 25 cm/14 in. × 14 in. × 10 in.). The roof can be hinged for easy access to the wasps, the bottom lower 2.5 cm (1 in.) of one side open and covered with 1 cm (⅜ in.) mesh wire to allow entry and exit of the wasps. (See Fig. 9.)

Boxes should be bolted to stakes and placed within or near cabbage plots. Although queen wasps may locate these sites and use them as nesting places the most reliable way to start a colony is to find a queen before or just after she emerges from hibernation and introduce her into the box in the hope that she will decide to use it to build her nest in.

As the nest is made of pulverized wood it is a good idea to have one or two untreated wooden posts or logs within the vicinity.

An isolated garden is ideal if wasps are going to be used as bio-control agents. In urban areas wasps may be a problem, particularly during the autumn. Some people are allergic to wasp stings. Wasps probably come under the nuisance legislation of the Public Health Act 1936, so great care must be employed when choosing sites for nest boxes.

Dragonflies

Dragonflies, by which I mean all members of the Odonata including damsel-flies, are opportunistic feeders, taking whatever prey is abundant. They therefore help to reduce pest insects but are unlikely to eradicate them.

Dragonflies will congregate when ants, mayflies, caddis-flies or gnats are swarming. Sometimes they appear to anticipate the presence of prey, arriving in the right place at the right time. Maturing dragonflies feed intensively, particularly the females when they are developing eggs.

Small dragonflies often perch on prominent sites while on the look-out for prey and they dash out when an insect flies past. In order to do this, they must keep warm enough by basking in the sun so that instant flight is possible. Such dragonflies therefore choose sites in the sun, while larger dragonflies normally feed by patrolling along bushes, hedges, woods and the like.

Adult dragonflies are sun loving and are not confined to the proximity of water. However, they must have water in which to breed. The aquatic nymphs are predatory and include nymphs of mayflies and dragonflies, tadpoles and even small fish in their diet.

Dragonflies lay their eggs within aquatic plants above or below the water, into the mud at the water's edge, on the stems or leaves of plants, or they scatter them over the water surface.

Bearing these points in mind, it is possible to encourage

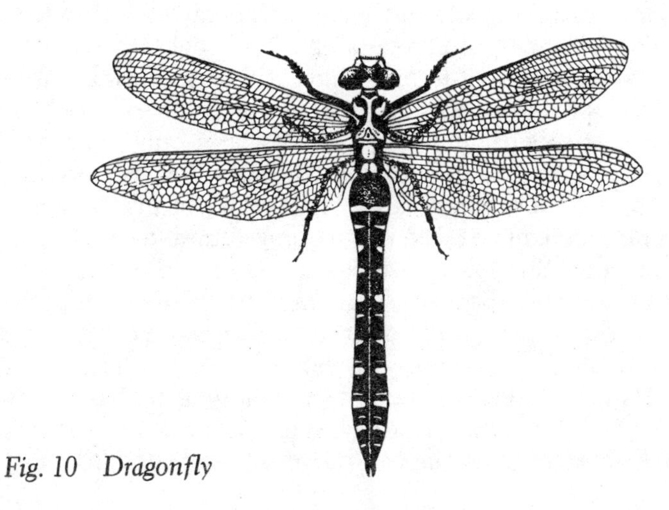

Fig. 10 Dragonfly

dragonflies to use your garden as a breeding site or patrolling site in order to catch prey. Creating a natural pond rather than an ornamental pond (see page 34) will encourage dragonflies to breed, while strategically placed posts will act as perches for patrolling dragonflies.

Bees

As well as bees, many insects visit flowers — hoverflies, beetles, flies and numerous others. However, it is bees — the solitary bees, bumble-bees and honey-bees — which are the most valuable and abundant flower pollinators. Honey-bees, in particular, nearly always keep to one species of flower during a foraging trip and in consequence increase the chances of the flowers being pollinated. Only where self-sterile varieties are present is fertilization a problem. But even then, bees do not always clean the pollen from their bodies thoroughly between foraging trips, so pollen collected on one trip may be used to cross-pollinate on the next. Pollen may also be transferred from bee to bee as they brush up against each other in the hive.

Cross-pollination is a particular problem in orchards. The best way round the problem is to plant each main variety of fruit tree which needs cross-pollination surrounded by pollinizer trees.

Bumble-bees are particularly valuable as pollinators, because they can work in low ambient temperatures when honey-bees and solitary bees are unable to work. Because of their large size, bumble-bees have considerable control over their body temperature, enabling their flight muscles to operate at low ambient temperatures. Whereas most bumble-bees are valuable as pollinators of open flowers such as blackberry, raspberry and fruit-tree blossoms, only those with long tongues are able to obtain nectar from flowers with long corolla tubes, such as red clover and runner beans.

Several steps can be taken to encourage bumble-bees to nest in your garden. Suitable nest sites and a continuous succession of flowers are both essential. An area of the garden can be set aside in order to plant wild-flower seeds. Cottage garden plants are particularly attractive to bumble-bees, whilst most members of the dead-nettle family including herbs, salvias, lavender, delphiniums, snapdragons and honeysuckle are also visited by them. Pussy-willow, white dead-nettle and flowering currant provide a food source for emerging queens in the spring.

Nesting sites and hibernation sites are best created by leaving an area of sheltered rough ground, with rotting wood, piles of vegetation (compost heaps are ideal), mounds of earth, piles of stone or stone walls, in which bumble-bees can build their nests. The queens choose a great variety of sites to build their nests in, but they all have one thing in common — they provide a dry, well-insulated home. This is why queen bumble-bees use the abandoned nests of other animals, as well as discarded artefacts of man's creation such as old armchairs and the like.

Perhaps one of the best ways to encourage bumble-bees is to provide nest boxes. These consist of untreated wooden rectangular boxes (35 cm×20 cm×12 cm/14 in.×8 in.×4¾ in.) with partitions and runners on either side of the floor to keep the nest box off the ground. An old roof tile provides a weatherproof roof. In early spring nesting can be encouraged by putting out nest boxes containing upholsterer's cotton-mouse bedding material which has become 'mousey' through use.

A very simple and easy way to provide a nesting site for bumble-bees is to use an earthenware flowerpot. (See Fig. 11.) Select a suitable site and dig a hole big enough to take a medium-sized flowerpot. Partially fill the pot with nesting material, and with the drain hole at the top place the pot into the ground so that the upturned base is level with the soil surface. Back fill with earth, leaving the upturned base exposed and cover this with a slate supported by pebbles so that the bees can crawl underneath and into the hole.

Solitary bees differ from social honey-bees and bumble-bees in that they do not live in communities but as individuals,

Fig. 11 A bumble-bee nest box

working alone and building and provisioning their nests for their young. Some solitary bees make their nests in burrows in the soil, while others build them in hollow stems of bramble, elder or decayed wood, using clay to close the cells. Some species may nest in close proximity to each other. Burrowing species are often attracted to sandy soil and may be encouraged by providing banks of sandy soil and growing early-flowering wild flowers such as dandelions or planting early-flowering aubretia.

It is possible to provide artificial homes for some species of solitary bees. Hollow bamboo canes approximately 15 cm (6 in.) in length will serve the purpose. Bundles held together with elastic bands can then be placed under the window-ledge of an outhouse or similar. Some species prefer to excavate their own tunnels and decayed logs placed in a warm corner of the garden provide the best conditions for success. However, beware of sun traps where temperatures can rise dramatically and kill the larvae.

Butterflies

Encouraging butterflies to breed in the garden is far from straightforward. One of the main problems is that just leaving the odd patch of food plants, many of which are pernicious weeds, in an odd corner is not enough. Butterflies are very fussy creatures. For instance, the small tortoiseshell butterfly only lays eggs on nettles growing in hot sunny positions and then only if they are young, succulent plants.

However, encouraging butterflies to visit the garden is another matter. Of the 58 species of butterfly that occur in Britain, sixteen regularly visit gardens, and although they have definite preferences most species will take nectar from a variety of flowers.

Butterflies are most numerous in the late summer, so is it important to plant flowers that flower at this time; although species which flower during the spring and autumn should also be included (see Appendix I). Both annuals and herbaceous plants should be planted in bold clumps so that there is a mass of flowers. One or two flowers by themselves do not lure butterflies.

Some butterflies, such as the small tortoiseshell and the peacock, overwinter as adults and appear on the first warm days of spring. The wall brown, green-veined white and small copper overwinter as either caterpillars or pupae and appear later, whereas immigrants such as the painted lady arrive in early June. Numbers then build up over the summer as butterflies breed, peaking in August and September.

Butterflies are generally attracted to common garden flowers, often preferring species plants rather than highly bred varieties. They also prefer sunny and sheltered spots — a south-facing border is an ideal spot for plants which attract butterflies.

Care must be taken when using the microbial insecticide *Bacillus thuringiensis* (or BT), as the pathogen is not only toxic to pest caterpillars such as the cabbage whites, but is also toxic to other butterfly caterpillars.

Other garden friends

Although this book is primarily about insects, a number of other animals play an important part in keeping down garden pests, in particular amphibians and reptiles, mammals such as hedgehogs and shrews, certain birds and garden creepy-crawlies such as centipedes, spiders and harvestmen.

Centipedes

Centipedes are predators, living on other invertebrates, and

Fig. 12 Centipede

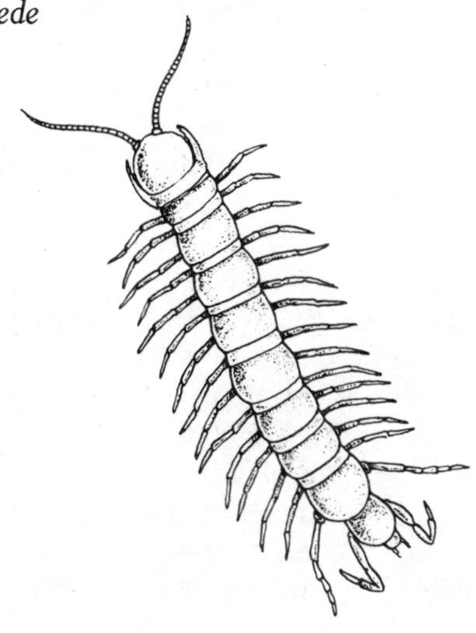

are able to tackle relatively large and active prey by means of their powerful poison-claws, with which they not only grasp but paralyse or even kill their victims outright. Unlike insects, centipedes have no waxy, waterproof cuticle so are obliged to live in moist surroundings; they also shun the daylight and avoid extremes of heat and cold. Their choice of microhabitat is largely governed by factors such as high humidity, a low light intensity and an even temperature. They are most likely found under stones, beneath the bark of decayed trees, between layers of decaying leaves and in the soil, compost heaps, beneath sacking and the like.

Centipedes are most often found in the spring and autumn when temperatures and humidity are ideal. During drought they retreat into crevices in the ground and during hard winter weather move deep into the ground to escape the cold.

Spiders

Spiders are particularly important predators. In China farmers provide straw 'teepees' in order to create shelter for overwintering spiders and consequently have reduced pest populations to acceptable levels.

Frogs and toads

These amphibians are particularly useful in keeping pests, such as slugs, in check. Frogs will usually breed in any convenient pond, but toads are more choosy and it is usually necessary to introduce some toad spawn into the garden pond in order to establish them in a garden.

The best way to construct an artificial but natural-looking garden pond is to use a butyl liner which will act as an impermeable layer. Butyl is very tough and durable and is less likely than PVC or polythene to be punctured by irregularities in the bedding or by accidental damage. The size and shape of the pond depends simply on the area of land available, your pocket and the preferences of the individual. However, the section of the pond demands certain requirements. In profile the pond should be saucer-shaped, approximately 75 cm (30 in.) or more deep, with the edges shelving away from the bank. (See Fig. 13.) It is important to consolidate the banks thoroughly so that there is no possibility of subsidence. Simply remove rocks, sharp stones, tree roots and other obstructions from the base and banks and cover the whole area with a 5 cm

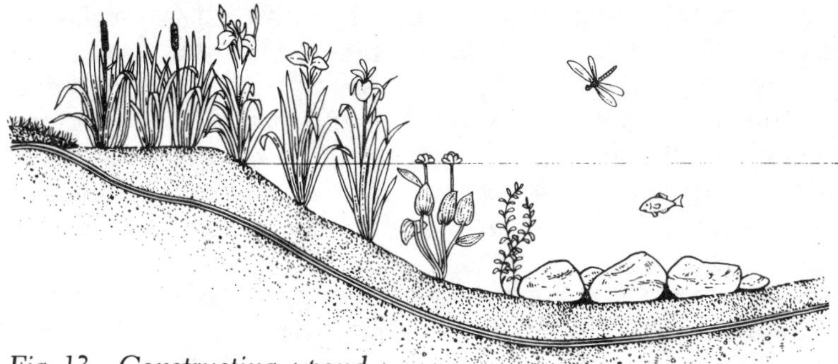

Fig. 13 Constructing a pond

(2 in.) layer of sand or polyester matting. On top of this, lay the butyl liner followed by another layer of polyester matting and 10 cm (4 in.) of washed, rounded pebbles or sandy subsoil. This impermeable sandwich can then be filled with water, planted with aquatic and marginal plants and the turf replaced up to the water's edges so that the grass remains fresh and green, and covers the liner edge. Within a few weeks the pond will have cleared, the grass grown back around the pond edges and it will look as though the pond had always been there.

Garden ponds can be death traps to small animals such as hedgehogs, and it is necessary to ensure that there are several slipways around the edge of the water. Ideally a gradual slope should be incorporated into part of the sides of the pond when it is built, which looks more natural in any case than ponds which are steeply sided all round. However, if this is not possible, rocks placed around the edge of the water will suffice.

Slow-worms

The favoured prey of slow-worms are slugs, but they will also eat spiders, worms and other soft-bodied insects and their larvae. These legless lizards are a great asset to the garden and are able to live in a wide range of habitats. They like warm places and often frequent sheltered sites such as compost heaps, piles of stones and the like. A sheltered, leaf-litter covered corner of the garden will provide an area for basking and hibernation.

Hedgehogs

The hedgehog is perhaps one of the best known gardener's friends as it eats slugs, beetles, caterpillars, millipedes, earwigs and other garden pests. Its benefits far outweigh its drawbacks. If you want to attract hedgehogs to your garden, leave wild areas and avoid 'tidying up' all those corners. Hedgehogs tend to hibernate between November and mid-March and may choose a pile of leaves or branches in a sheltered spot of the garden in which to hibernate. They also like to nest under things such as sheds or hedges and need plenty of dry leaves for the purpose.

Do not be tempted to give hedgehogs bread and milk, as micro-organisms found in cow's milk do not agree with them. Feed them on tinned meat used for dogs or cats, if you want to feed them.

Providing shelter and food

Bats, hedgehogs and some birds will adopt artificial boxes in which to roost, nest or hibernate. These boxes can be purchased, but are quite easy to make (See Fig. 14.) Bats prefer a rough surface on which to alight and roost, so their boxes should be constructed from rough-sawn timber. All types of boxes should be left untreated, as the smell of preservatives repels most species. Position greatly influences their attraction. Bat boxes are best placed under the house eaves or fixed to trees at different heights (at least 1.5 m/5 ft off the

Fig. 14 Constructing a bat box

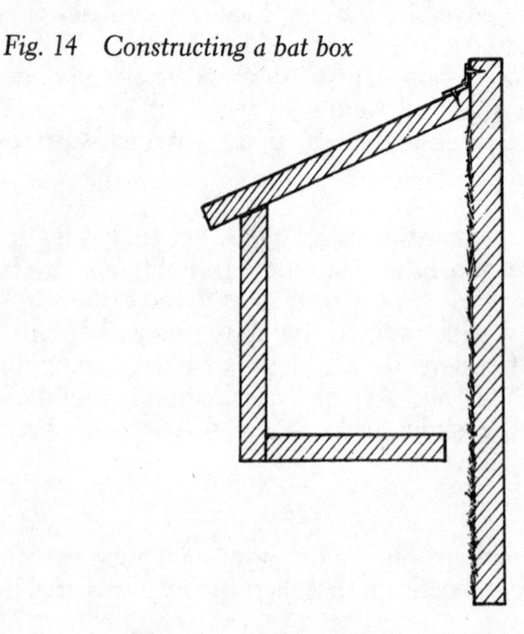

ground) and aligned in different directions. South-east facing boxes are preferred in summer while a northerly aspect is preferred for hibernation; but be prepared to wait a long time — maybe several years — before a bat will take up residence.

Hedgehog boxes should be placed in a quiet sheltered spot and provided with dry leaves or hay which can be used for bedding.

Bird boxes should be attached to a piece of wood which is screwed to either a tree or a post. This will prevent rain water trickling down the back and causing the box to get damp. Position in terms of sun, rain or wind is most important. The prevailing wind will cool down the nest box considerably and also drive rain into the hole. Sheltered sites should be selected but there must be a clear flight path to and from the box. Boxes should also be placed at a distance from each other, as although birds will tolerate other species to some degree they will not tolerate intrusion from their own kind.

Many common birds are insectivorous but some — such as the blackbird — eat a wide range of food, including invertebrates, seeds, berries and fruit (see Appendix I). An anvil in the form of strategically placed rocks will provide a place for thrushes to break open snail shells, in order to get at the tissues inside.

Birds should be provided with food and water during severe winter weather. Various dispensers are available, some of which exclude all kinds of birds except the titmouse. Whilst tits are lining up on nearby trees and shrubs they will search for any overwintering pests in the vicinity. It is also a good idea to plant shrubs and other plants which will provide food, usually in the form of berries, for the birds (see Appendix I).

3
Garden foes and how to control them

What are pests?

A garden pest or foe can be defined as an animal, usually the creepy-crawly kind, that causes damage to cultivated plants by its feeding or living habits. Whether or not this justifies control measures depends on your point of view.

You do not need to control all pests all of the time. Some pests cause relatively little damage and the benefits of using a pesticide or other control method are usually quite small. It is the pests which significantly affect the yield or appearance of the plant that need controlling. If left unchecked these pests will distort and stunt plant growth and generally spoil appearance. Pests can be loosely divided into three groups, depending on the type of damage they do to plants.

Sap feeders

Insects such as aphids, whiteflies, scale insects, thrips, mealy-bugs, leaf-hoppers, frog-hoppers and capsid bugs, and mites such as the red spider mites, have needle-like mouthparts which they use to pierce the plant and suck the sap. These pests reduce vigour of plants and may cause distortion of leaves and buds. Usually sap-feeding pests do not make visible holes in the plant, but the saliva of capsid bugs causes holes to develop in affected leaves whereas leaf-hoppers, thrips and red spider mites may cause the plant to lose its greenness.

Aphids, scale insects, mealy-bugs and whiteflies feed from the veins of leaves and stems. The sap is so rich in sugars that there is too much for them to digest, so they excrete the surplus as honeydew. This makes the leaves of the plant very sticky which not only attracts ants but encourages a black fungus called sooty mould to grow. Sooty moulds look unsightly and prevent light from reaching the leaves. It is common on deciduous trees during the summer. Lime in particular, but also oak, willow, sycamore and birch become heavily infested with aphids and their honeydew which affects not only their foliage but drips onto other plants growing below. Aphids and other sap feeders can cause further damage by spreading virus diseases. Viruses are taken up with the sap from an infected plant and are carried on the insect's mouth-parts or in the saliva to healthy plants.

Foliage, root and stem feeders

Caterpillars, beetles, earwigs and other creepy-crawlies like woodlice, millipedes and slugs, use their mouthparts to cut or rasp holes in the leaves, roots, stems and flowers of plants. Often it is the larval stage that does the damage rather than the adult.

The extent of the damage can vary from a nibbling of the edges of leaves to complete defoliation of the plant. Some sawfly and beetle larvae only eat the lower or upper surface of the leaf which causes the damaged area to dry up, reducing photosynthesis.

Root feeders can cause considerable damage to the roots of plants and usually go unnoticed until the plant suddenly wilts and often dies as a result.

Some pests feed inside the stem, roots or leaves of plants. Leaf miners, for instance, actively tunnel through the leaves which clearly show as white or brown trails or patches against the green of the leaf surface. Leaf miners are in fact the collective term used for the larvae of various sawflies, beetles and moths. The shape and size of the mine can be used to identify the pest responsible.

Gall-forming pests

A gall is formed in response to attack by a pest or disease. A wide range of pests can stimulate a plant to form a gall, including aphids, sawflies, gall-midges and gall-wasps. Each species usually affects only one particular type of plant. Galls are relatively harmless to plants and very rarely need controlling. Some gall-forming pests, however, can cause damage if they occur in large numbers. Galls occur in various forms: a hairiness on the undersides of leaves, discs or tubular growths on the leaves, enlarged buds or a thickening and curling of leaf edges.

Controlling pests

In nature, pests are usually kept in check by their natural enemies. If you create a garden that has a variety of both plant and animal life, the more sustainable it is likely to be. Diversity is the all-important word. To create the right balance it is important to plant native trees and flowers; devote part of the garden for this purpose and for creating a pond where wildlife can flourish.

Only when pests build up to unacceptable levels should pesticides be used. Consider using pesticides only when yields of fruit or vegetables are likely to be greatly reduced; when plants suffer from the same pests and diseases every year and where there is likely to be a spread of disease to other plants, especially virus diseases. There are, however, other ways to prevent pests and diseases without recourse to the use of pesticides.

A great deal can be done initially to help plants withstand attacks by pests and to help prevent the establishment of disease in the garden. For instance, cultivate the soil thoroughly, especially in the winter, so that pests are exposed to frost and birds and in consequence the numbers that survive are reduced. It is also possible to time seed sowing in order to avoid the worst attacks by pests. You should water and feed plants to encourage steady growth but do not overfeed them, especially with high nitrogen fertilizers, as this will encourage soft sappy growth which encourages aphids and other sap feeders. It is important to make sure the condition of the soil is good at planting or sowing time in order to ensure a rapid start in life for seeds or plants so that they can withstand initial attacks by pests and diseases, such as damping-off disease.

Keep the vegetable garden tidy by removing and burning diseased plants and vegetation; this in turn helps remove potential sources of infection and hiding places for many pests. It is important to keep down weeds which act as alternative hosts for many pests; for instance, weeds such as shepherd's purse, charlock and wild radish (Cruciferae) may act as host plants for carrot fly and flea beetles. Dense weed-cover maintains damp conditions at soil level which favour pests such as slugs, woodlice and millipedes. Plants which become heavily infested with pests should be removed and burnt before the pests spread to nearby plants. Do not put infested or diseased plants on the compost heap because many spores of fungi and pests, such as wireworms, can withstand composting and may even multiply.

There are now many varieties of plants which are resistant to pests or diseases or differ in their susceptibility. Those varieties which consistently succumb to disease or pest damage should be avoided and replaced with a variety which is resistant (see Appendix II).

Removing individual pests by hand may be a time-consuming business but can prove very effective in small gardens or where few plants such as cabbages are affected.

There is considerable anecdotal evidence to suggest that certain plant associations deter or inhibit pests. However, there is little scientific evidence to substantiate the claims. All the plant associations mentioned in this book have a sound scientific basis. (See *intercropping* sections for specific pests.)

Traps and barriers are also useful for keeping down overall populations of pests. Brassica collars around cabbages are particularly effective in preventing cabbage root fly, whereas plastic barriers around carrots help stop carrot fly. (For details, see specific pests.)

Biological control

The use of biological control methods can be important, especially in the controlled environment of the greenhouse. Even outside slugs can be controlled by the use of ground beetles, while the white butterfly's caterpillars can be controlled by using the microbial insecticide *Bacillus thuringiensis* (BT), a bacterium used in solution as a spray.

Insecticides derived from plants and natural products

There is a common assumption that using a spray made from natural products such as the roots, bark or flowers of plants is a safe way to control pests. All pesticides are toxic; some are more toxic than others. The advantage of pesticides made from natural products is their low toxicity to birds and mammals when diluted and the fact that they are rapidly biodegradable and do not accumulate in the environment. However, they are often toxic to fish and useful insects. It is always advisable to wear a mask and gloves during application.

Nicotine

Nicotine is a tobacco extract highly toxic to warm-blooded animals and has been withdrawn from sale to the general public, although it is still available to professional growers.

As a commercial insecticide it usually consists of a liquid concentrate of nicotine sulphate, which is diluted with water and applied as a spray. Dusts can irritate the skin and are not normally available for garden use. My grandfather John Moffat Forsythe would have made his own nicotine insecticide by soaking cigarette tobacco in water and using the concentrate as a spray or in a fumigator in the greenhouse overnight. Nicotine is used primarily for piercing-sucking insects such as aphids, whiteflies, thrips and leaf-hoppers. It is more effective when applied during warm weather and because it is rapidly biodegradable can safely be used on plants nearing harvest.

It is harmless to the eggs and larvae of the two- and seven-spot ladybird, adult seven-spot ladybirds, hoverfly larvae, adult and larvae of the green lacewing and several predatory mites. It has only a slight effect on predatory capsid and anthorcorid bugs and is only slightly toxic to parasitic wasps, but it is toxic to bees, so should only be used in the evenings. Nicotine sulphate biodegrades rapidly and has little residual effects.

Derris

Derris is a non-persistent insecticide extracted from the roots of the derris species of plants. Its active principle is rotenone and it is used either as a dust or spray. It has a very low toxicity to warm-blooded animals and leaves no harmful residue on vegetable crops. However, it is very toxic to fish. It is used for the control of flea beetles, raspberry beetles, asparagus beetles, aphids, caterpillars and sawfly caterpillars.

Derris is harmful to beneficial insects; it is toxic to the eggs, larvae and to a lesser extent the adults of ladybird beetles and is highly toxic to adult green lacewings, some parasitic wasps and predatory mites and to the honey-bee. Hoverfly larvae are unaffected and toxicity to the green lacewing larva is low. It acts as both a contact and stomach poison to insects.

Ryania

Ryania is an insecticide extracted from the wood of *Ryania speciosa*. Its active principle is ryanodine and it is used primarily for the control of codling moth, although it appears only moderately effective in the British Isles.

Ryania has a low toxicity to warm-blooded animals and most beneficial insects and predatory mites. It acts as both a contact and stomach poison to insects.

Quassia

Quassia is a pesticide extracted from the Simaroubaceae family of plants. Species of *Balanites, Picrasma* and *Quassia* are all used.

Quassia usually comes as chips of wood which can be boiled up in water and used as a spray against aphids, sawfly caterpillars and mites. It is harmless to bees and ladybird beetles.

Sabadilla

Sabadilla is an insecticidal stomach poison for use against insects from the plant *Schoenocaulon officinale.* It is highly toxic to honey-bees and to mammals and is slightly toxic to lacewings. Sabadilla rapidly deteriorates in the presence of light, so it should be applied during late evening which will minimize undesirable effects on bees.

Sabadilla is used in the USA as a general insecticide against a wide range of pests. At present not available in the UK, it may become so in the near future.

Pyrethrum

Pyrethrum is prepared mainly from the flower heads of *Chrysanthemum cinerariaefolium,* which has several insecticidal components known as pyrethrins. These should not be confused with pyrethroids which are synthetically produced.

Pyrethrum decomposes rapidly, especially in sunlight, and therefore has a short residual effect. It is toxic to fly pests, aphids, thrips, whiteflies and other insect pests. It is also toxic to ladybirds, some parasitic wasps and fish.

Its unique 'knock down' effect stuns flying insects, quickly causing them to fall to the ground. Because it is irritating, pyrethrum flushes insects out of their hiding places, thus exposing them to the elements and predators and brings them into contact with the spray.

Pyrethrum is usually incorporated with a synergist *Piperonyl butoxide* which is extracted from a species of *Sassafras* and increases the effectiveness of pyrethrins.

It can be used up until the day before harvest.

Insecticidal soap

Insecticidal soaps are used to kill aphids, mealy-bugs, white-flies, mites, sawfly caterpillars and other pests. They are not toxic to honey-bees and beneficial insects, and can be used on all crops right up to harvest. They do not harm wildlife and are completely biodegradable.

Insecticidal soaps are specially formulated salts of fatty acids and are used in dilution with water (rain-water) as sprays.

It is not clear exactly how insecticidal soaps kill pests but it appears they do so in several different ways: by paralysis of the nervous system, breakdown of the haemolymph (blood), disruption of respiration and the waxy coating on the insect's body, disrupting its osmotic balance.

Insecticidal soap sprays are contact insecticides so it is important to spray both the top and underside of the leaves. They have been tested on a wide range of vegetable crops and fruit without any phytotoxicity problems, but to be on the safe side, a small-scale testing on plants is advised before engaging in full-scale spray application.

It should be pointed out, however, that some ornamental plants are susceptible to insecticidal soaps.

Insecticidal soaps can be applied with simple spray equipment and do not require use of special safety gear. Spray early in the morning, or in the evening, or when the sky is overcast to prevent foliage discoloration as a result of sunlight shining through droplets of moisture on the leaves of plants. Insecticidal soap should not be confused with household soap or detergent which are manufactured using a different process and are not recommended for garden use.

Soft soap

Soft soap makes an excellent wetting agent and is often used with plant-derived pesticides, such as quassia, to improve their effectiveness. It is also mildly insecticidal in its own right and is effective against aphids.

Proprietary insecticides

Details are given in Appendix II.

Microbial insecticide

The bacterium *Bacillus thuringiensis* (BT) has been commercially available as a microbial insecticide since 1971. It is selectively toxic to Lepidopterous caterpillars and has been used successfully against over 100 species of caterpillars on a wide range of non-edible and edible crops. BT is naturally occurring, is highly specific in activity and is non-toxic to animals and beneficial insects.

BT has a two-pronged attack. Not only are the spores themselves infective, but BT contains a crystalline structure which is produced in association with each spore. When broken down in the gut of the caterpillar the degradation products are active against the gut lining, causing small lesions and the insect stops feeding, usually within 30 minutes of receiving a dose. The mixing causes severe disruption of the metabolism and death follows within about 4–5 days.

Potato Starch

A contact insecticide made from potato carbohydrate and effective against aphids, plant bugs and mites. Currently being tested by the UK Ministry of Agriculture.

Moths

Codling moths

The maggots found in apples and pears are usually those of the codling moths. They lay their eggs on the developing fruit in June and July and the larvae burrow inside and feed on the central core. By the end of August most of them will have eaten their way out of the fruit, leaving an exit hole on the outside and 'frass' (excrement) in the apple. If there is a second generation of moths, the maggots may still be in the fruit when harvested.

Fig. 15 Codling moth

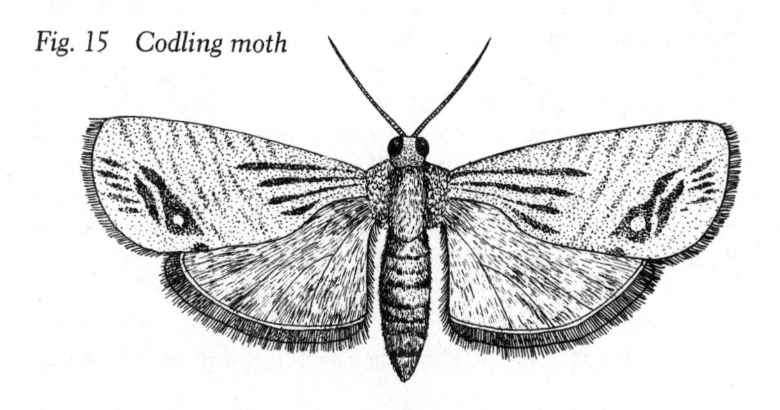

Fig. 16 Codling moth trap

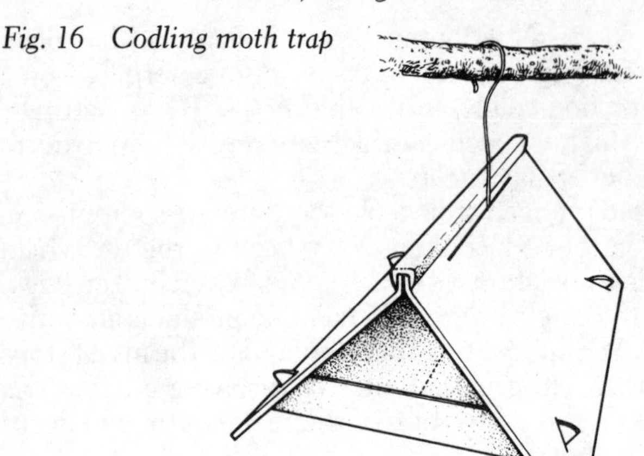

Traps

Traditionally pieces of sacking or corrugated cardboard were tied around the trunk and larger branches of fruit trees.

These traps attract larvae as they move down the tree to overwinter under bark or ground cover or other suitable shelter. Occupied traps can then be destroyed by burning.

In recent times pheromone traps have been developed to monitor codling moth populations. These traps are, however, unlikely to affect population levels although each trap will catch many hundreds of moths during a season. They really only determine whether your apple trees need spraying, and when to spray.

The traps contain an artificial version of the scent (pheromone) that female codling moths use to attract males. Thinking that they have located a mate the male moths fly to the lure and get caught on a non-drying sticky adhesive.

There are two types of trap: the delta trap and, for larger moths, the funnel trap. Both are 'self assembly', and both contain a piece of pheromone-impregnated rubber. The delta trap also contains a non-drying sticky adhesive and the funnel trap a toxicant. Both traps are hung in fruit trees and are used to monitor the populations of codling moths present in the orchard. (See Appendix III for suppliers.)

Natural insecticide

The above traps give instructions on when and when not to

spray and are mainly for use with synthetic pesticides. It is possible, however, to achieve limited control by substituting ryania, a plant-derived pesticide, which is not very toxic to mammals but a stomach poison to plant-feeding insects.

The microbial insecticide *Bacillus thurungiensis* (BT) is effective against codling moths, but unfortunately the larvae are tucked away inside the apple and are unlikely to come into contact with the pathogen.

Other moth pests

Formulated lures are available for monitoring other moth pests. They work on the same principle as the codling moth trap described above. Some of the more common moth pests for which lures are available are listed below.

Leek moth *Acrolepiopsis assectella*
Cutworm *Agrotis exclamationis*
Turnip moth *Agrotis segetum*
Pea moth *Cydia nigricana*
Cabbage moth *Mamestra brassicae*
Winter moth *Operophtera brumata*
Diamond-back moth *Plutella xylostella*
Current clearwing moth *Synanthedon tipuliformis*
Fruit-tree Tortrix moth *Archips podana*
Summer-fruit Tortrix moth *Adoxyophyes orana*
Fruitlet-mining Tortrix *Pammene rhediella*
Plum moth *Grapholitha funebrana*

Pea moth

Pea moths emerge from cocoons in the soil from early June onwards and are most active on warm, sunny days in July. The

Fig. 17 Cabbage moth

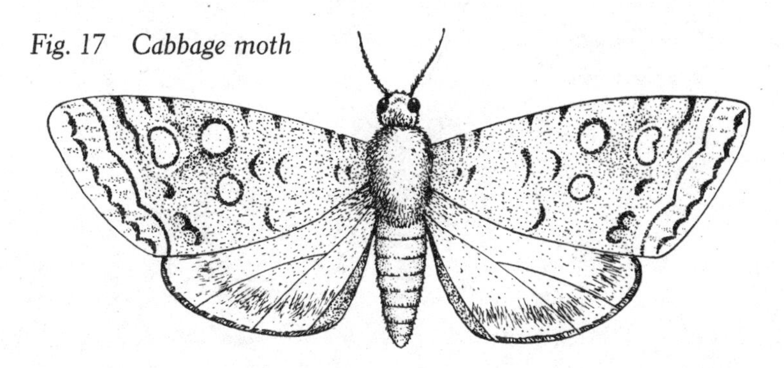

female moth lays her eggs in pea plants and the caterpillars eat their way into the young pea pods and feed off the developing peas. They are easily recognized as small, pale-yellow caterpillars with black heads.

Pea moth hosts include not only garden varieties of leguminous plants, such as garden peas and sweet peas, but also wild species such as meadow vetchling and tufted vetch.

The caterpillars (grubs) are protected from chemical sprays as they are inside the pods for most of the time.

Cultural control

Peas from crops that flower during the flight period (early June to mid-August) suffer most damage. Egg laying is mainly restricted to June and July, so early or late sowing can reduce serious infestations by avoiding the flowering and pod formation. Do not sow peas in March or April, but sow earlier or later in the year.

Winter cultivation may also help by exposing cocoons to the weather and predators.

Natural enemies

The natural enemies of the pea moth are four species of hymenopterous parasites.

Traps

As specified above.

Winter moths

The female winter moth is wingless and emerges from its pupa

Fig. 18 Winter moth

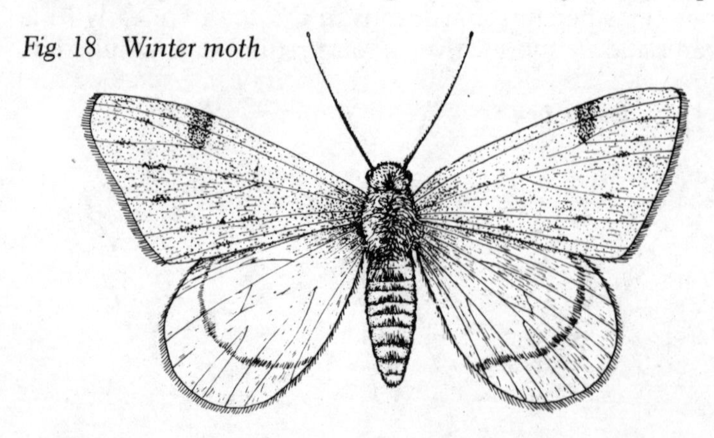

in the soil sometime during the period October to December. It mates with the winged male and then crawls up the trunks of fruit trees or ornamental trees to lay its eggs in small clusters near buds and in crevices in the bark.

The caterpillars finish feeding sometime between mid-May and the end of June when they move down the tree to pupate in the soil. The pupae remain dormant in the soil until they emerge as adult moths in the winter.

Barriers

Grease bands are effective against wingless females of the winter moth and prevent them from climbing the trees to lay their eggs. They should be in place from October to February. Tie bands about 10cm (4 in.) wide 1–2m (3–6 ft) above soil level around the trunks after removing any loose or rough bark and make sure that female moths cannot bypass the band by crawling over it on overhanging branches which come into contact with the trunk or ground.

Natural insecticide

Derris — spray trees immediately after buds open in the spring.

The microbial insecticide *Bacillus thuringiensis* is also effective against the winter moth.

Tar oil

Tar oil, a by-product of the petroleum industry, is used as a winter wash on dormant fruit trees and bushes and will give some protection against winter moth by killing overwintering eggs.

Two other winter moths, the march moth *Alsophila aescularia* and mottled umber moth *Erannis defoliaria* can be controlled in the same way as the winter moth.

Natural enemies

The forest bug *Pentatoma rufipes* overwinters as a nymph and in the spring preys on the caterpillars of the winter moth. The burying beetle *Dendroxena quadripunctata* has an unusual habit for a burying beetle since adult beetles climb bushes and trees to hunt caterpillars, particularly of the winter moth and mottled umber moth.

Species of tachinid fly and ichneumon fly are parasitic on

Fig. 19 Large white butterfly

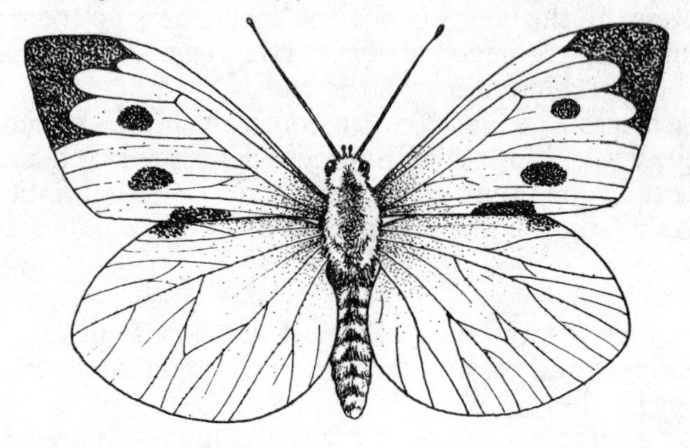

winter moths and some species of ground beetle are predatory also.

Cabbage caterpillars

There are six species of butterflies and moths which commonly attack brassica crops, four of which are widespread in the UK.

Caterpillar damage often encourages secondary rotting and plants are often fouled with the 'frass' produced by caterpillars. Some species feed on the outer leaves, others feed on the inner leaves, while other species feed under a silken web.

The most serious damage is done to mature plants in the summer, autumn and early winter.

The three species most likely to cause damage are the large white butterfly *Pieris brassicae*, the small white butterfly *Pieris rapae* and the cabbage moth *Mamestra brassicae*. Both butterflies have two or three generations per year whereas the cabbage moth has just two. Adults are to be seen from February to October, depending on species. Occasionally, large immigrations of diamond-back moth *Plutella xylostella* from the Continent cause damage also.

Traps

Use those for cabbage moth and diamond-back moth (see page 46).

Natural insecticide

Derris.

Intercropping

The presence of ground cover provides a habitat which is more suitable for some of the predators of the cabbage butterflies and also visually masks the crop. The presence of weeds or red or white clover help to reduce the numbers of caterpillars on brassica crops. However, due to the competition, yields in weedy plots are lower.

Successful intercropping trials conducted in the Philippines have resulted in a reduction in abundance and egg laying of the diamond-back moth. It seems a cabbage–tomato intercrop of two rows of cabbage between two rows of tomatoes is the most effective, as the tomato odour masks that of the cabbage. The results may not be valid in UK conditions, but may be worth trying.

Natural enemies

Cabbage caterpillars have numerous natural enemies. Predation by birds is important, particularly starlings which destroy large numbers of caterpillars. Other predators of young caterpillars include anthocorid bugs, hoverfly larvae, spiders, harvestmen and several species of ground beetle.

Microbial insecticide

Some of the more common lepidoptern pests susceptible to *Bacillus thuringiensis* are listed below:

Vegetables

Diamond-back moth *Plutella xylostella*
Cabbage moth *Mamestra brassicae*
Small white *Pieris rapae*
Large white *Pieris brassicae*

Fruit trees

Leafrollers *Archips* spp.
Lackey or tent caterpillars *Malacosoma* spp.
Winter moth *Operophtera brumata*
Fruit-tree Tortrix moth *Archips podana*

Cutworms

Cutworms are the caterpillars of certain moths which emerge at night and severely damage plants by cutting them off at ground level. The commonest species are the turnip moth,

Agrotis segetum; yellow underwing, *Noctua pronuba*; the garden dart moth *Euxoa nigricans* and the white-line dart moth *Euxoa tritici*. Most damage to crops is done in July and August.

Natural enemies
Predatory beetles such as ground beetles and rove beetles.

Cultural control
Cutworms are most common on weedy land, since it provides the cover attractive to egg-laying moths and food for the caterpillars. Land should be cleared of weeds well before the crop is planted and kept free of weeds during the life of the crop.

Young caterpillars, especially those of the turnip moth, die in wet soil. Frequent watering during July and August should prevent infestation.

Natural insecticide
Caterpillars live in the upper few centimetres of the soil and feed on the surface mainly at night, and are therefore difficult to control with short-lived, non-persistent insecticides like derris. It is perhaps best to encourage nocturnal predators such as ground beetles and rove beetles.

Traps
See page 46.

Slugs

Slugs are perhaps one of the most insidious of pests and very difficult to control without using chemicals.

If control of slugs is to be effective it is important to understand something of their ecology and behaviour. Slugs are active and feed throughout the year whenever the temperature and humidity conditions are suitable. During very dry or frosty weather they stop feeding and move down into the soil or shelter under stones or debris. They are most active at night when the soil is wet and the atmosphere humid. They are least active during heavy rain or on windy nights. A damp environment is essential to them as slugs breathe partially through their skins.

The effect of adverse weather conditions then is clear.

When shelter is scarce large numbers of slugs are killed by drought or frost. Weather can also have a less direct effect on slug populations — by shortening or lengthening the generation interval. This may be more important in determining slug density than the direct effect of weather on mortality.

The effect of predators and parasites on populations of slugs may also be determined by the quantity and type of available shelter.

It seems, therefore, that there are three key factors which affect slug populations: firstly, the action of predators; secondly, the direct effect of frost and drought on mortality; thirdly, the indirect effect of weather conditions on generation interval.

Slugs feed on decaying plant and animal material, as well as living plant material such as seedlings and flowers. They also feed on the crowns and roots of many plants and underground tubers such as potatoes.

Because they like to eat rotting plant material they are often to be seen in large numbers in compost heaps. Here they shelter and breed and in doing so help to break down organic matter. In this way they can be regarded as beneficial.

Although there are many species of slugs in the British Isles only three kinds generally do serious damage to crops:

1 *The field slug* (*Agriolimax reticulatus*) is variable in colour, but is usually light grey or fawn, measuring about 2.5cm (1 in.) long and has a milky mucus coat. This is probably the most common and most destructive slug. It is active at low temperatures and feeds above ground on the aerial parts of plants, below ground and on the surface.
2 *The garden slug* (*Arion hortensis*) is black on its upper surface and yellow underneath with a slightly yellow mucus coat. It is slightly smaller than the field slug. It is a destructive and costly pest, and feeds above and below ground level.
3 *The keeled slug* (*Milax* spp.) is grey, dark brown or black with

Fig. 20 Slug

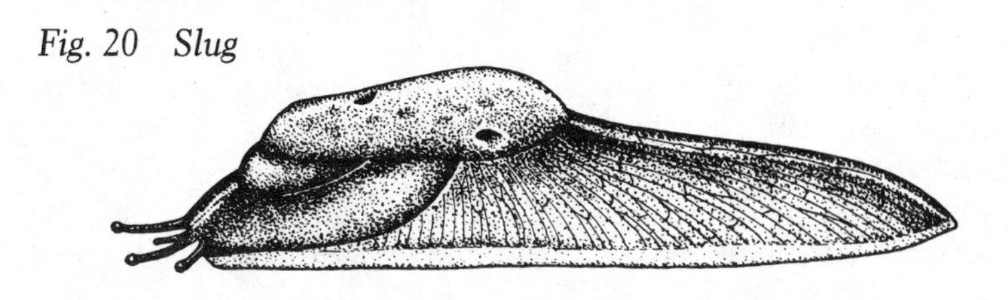

a ridge or keel running down the centre of the back. It is larger than the two previous species. It is largely subterranean but will also feed above ground. In some areas it is an important pest.

Chemical control

None of the following chemical methods of controlling slugs is recommended if a natural organic environment is desired. Nevertheless it is important to know how their use affects the ecosystem. The two compounds containing aluminium sulphate are often on sale as slug killers in organic gardening centres under various brand names, probably because they are claimed to be harmless to wildlife. Read the labels on the pack carefully.

Metaldehyde

This works either as a contact poison or stomach poison. The molluscicidal effect is based on two principles: firstly, its irritant effect causes the slugs to secrete large amounts of mucus which results in desiccation and death: secondly, its toxicity as a nerve poison prevents the slug crawling away from the poison. Metaldehyde is usually presented incorporated into a pellet with a bait such as bran; as a liquid concentrate: as a cream containing an attractant: as loose bran or incorporated into a tape coated with a bait (sugar).

Slug-tape is a paper-based tape. The molluscicide is safely sealed inside a protective outer layer which is free from harmful ingredients and coated with a non-toxic attractant appetizing to slugs but not to animals. It has a very slow decay rate and remains active for over eighteen weeks in all conditions. It is laid flat or in a collar around individual plants.

These products usually contain a blue dye which is supposed to discourage birds from eating them. They also contain animal repellants to deter cats and dogs.

A similar substance *methiocarb* which is more toxic to slugs than metaldehyde is also incorporated into pellets. It is also more toxic to pets and wildlife and can kill beneficial insects, too.

Both metaldehyde and methiocarb are dangerous to animals, birds, fish, bees and other insects.

Aluminium sulphate

This is a white powder which is sprinkled onto the soil or dissolved and watered in. It aims at killing slugs above and below the soil surface and is reputed to be harmless to animals other than slugs and snails.

Aluminium sulphate, copper sulphate, borax mixture

This is a wettable powder which is sprayed or watered onto the soil. It is claimed to work by leaching down into the soil to egg-laying depth so that as the slugs emerge they are killed. It is reputed to be harmless to plant life, insects and worms.

In acid soils the availability of aluminium increases markedly and toxicity can be a major cause of crop failure. Symptoms of aluminium toxicity include slow growth, dull purple leaves and root discoloration. Crops most susceptible include celery, beet and lettuce. The dosage rate recommended for these products is probably too low to cause any lasting damage, and the aluminium in the presented state may not be available to plants.

Baits

Baits are used to lure slugs away from vulnerable plants to one spot where they can be collected and destroyed. Grapefruit skins, potato shells, shredded carrot, cabbage peelings, lettuce, banana peel, bran, wheat products and yeast can all be used to attract slugs. They should be inspected daily and the slugs removed.

Using baits to attract slugs can present a number of problems. For instance, the food preferences of slugs in a particular area can change with time; slugs tend also to lose interest in the same kind of bait. The type and amount of other available food in the area influences the attractiveness of the bait; if the food is juicier than the bait the attractiveness of the latter is lost. On the other hand, if the food is drier than the bait, the bait remains attractive.

Baits containing mainly animal protein are not recommended because they tend to improve the condition of slugs and their resistance to parasites and disease.

Traps

Slugs are attracted to alcohol. This response can be used most effectively by using alcohol, such as beer, in some form of trap. A saucer, bottle or other receptacle buried in the ground up to the lip and filled with beer should attract and drown slugs. At least they die happy!

It has the major disadvantage of also capturing ground

beetles, shrews and other predators.

Slugs seek shelter during the day in cool, damp areas — under bricks, large stones, damp sacking, wooden planks and the like. These shelters can be turned over (and replaced) every day and the slugs picked off.

Barriers

Many types of barrier which physically prevent slugs from reaching vulnerable plants have been tried.

Ammonium sulphamate, salt, soot, wood ashes, slaked lime, builder's plaster, all act as chemical barriers, but usually get washed away and are often less effective when wet — and when conditions for slugs are ideal.

Also some of these materials are harmful to plants (ammonium sulphamate) or to the soil if used to excess (salt).

Sharp sand, crushed eggshells and similar sharp materials also act as physical barriers to slugs.

From my experience none of the above methods have proved very effective.

Mulches

There are naturally occurring substances which kill molluscs or which reduce their fecundity. Many plant extracts have molluscicidal properties, some of which depend on the presence of saponins and phenolic compounds, such as tannins.

Tannins are present in oak leaves and oak bark. A thick mulch of composted oak leaves or tan bark (used in the tanning industry) may help repel slugs. Alternatively, a mulch of composted pine bark may have similar results. Remember an uncomposted mulch may *attract* slugs rather than detract them.

Saponins are present in plants such as soapwort and quinoa. A dried powdered preparation may also help to repel slugs. The problem with any powdered product is that it is totally ineffective in wet weather — the most prodigious time for slugs.

Cultural methods

It has already been noted that shelter is an important factor in the establishment and maintenance of slug populations. Therefore, if you want to reduce these populations it is important to remove any garden debris left over from the previous crop. Thorough cultivation of the soil breaks up clods of earth and reduces large cracks and spaces in the soil so that slugs cannot find shelter from drought, frost or predators. Also without such spaces, the slugs are unable to move through the soil to reach growing crops. Careful weed control reduces the vegetation at soil level, allowing the soil surface to become drier with an overall reduction in humidity levels which will be less favourable to slugs. When you do water, do so long enough to wet the deeper layers of the soil. This way plant roots will move downwards into the soil and be less susceptible to drying or other surface damage. It is important to improve the soil by adding well-composted organic matter. This will improve the chances of crops surviving slug attacks. It may be interesting to note that ploughing in straw to improve the soil structure leads not only to an increase in slug damage but also to nitrogen deficiency. (Perhaps this is why farmers burn their stubble.)

Another cultural method of reducing damage by slugs is to grow varieties of crops that are less susceptible to attack (see Appendix II).

Natural enemies

As slugs are ubiquitous, slow moving and available throughout the year they potentially provide endless scope for exploitation by predators, parasites and micro-organisms.

Slugs are regularly eaten by birds such as starlings, jackdaws, rooks and gulls. Most of these birds forage in flocks and probably make significant impact on slug populations.

Hedgehogs, shrews, frogs, toads, newts and slow-worms also eat slugs.

The main insect predators of slugs are beetles belonging to the families Carabidae (ground beetles), Staphylinidae (devil's coach-horse) and Lampyridae (glow-worm), and flies belonging to the family Sciomyzidae. There are 65 species of Sciomyzidae in Britain and the entire family specializes in attacking live molluscs. They are better described as parasitoids because the young larva penetrates the host and becomes an internal parasite. When the host mollusc dies the larva eats the remains and moves off in search of further hosts which it consumes as a predator. Most species live off aquatic snails but a few feed on terrestrial snails and slugs.

Ground beetles are probably the most important insect predators of molluscs. Some of the larger species, such as the violate ground beetles and their relatives, feed almost exclusively on soft-bodied organisms such as slugs and snails. The larvae are also predominantly predatory and may be even more beneficial than the adults. This predatory habit can be utilized in order to reduce slug populations. Plants vulnerable to slug damage can be planted in what is in reality a giant pitfall trap which allows ground beetles to fall in but does not allow them to climb out. (This idea has been developed by Bill Symondson at Cardiff University.) Once in, the beetles will feed on any slugs living inside the trap and in doing so reduce the slug populations to acceptable levels. Ground beetles, like slugs, are mainly nocturnal so any underground slugs which appear in order to feed at night will be eaten. Some ground beetles will even climb plants in order to find prey so that slugs feeding on the aerial parts of plants will not escape.

Making a pitfall trap
To build a trap dig a trench approximately 13cm (5 in.) deep and 1m (1 yd) square, with the outside edges vertical and the floor sloping gradually towards the centre. Board the vertical edge and run a length of lawn edging all the way round, holding it in place with pieces of wire and back fill with about 2.5cm (1 in.) of soil. This will give a smooth vertical wall which will prevent the beetles from escaping. (See Fig. 21.)

Fig. 21 Pitfall plot

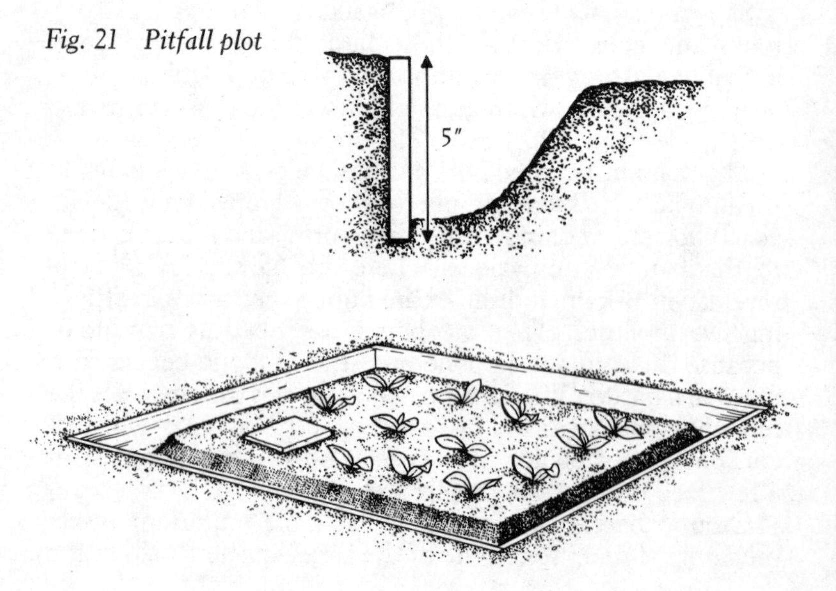

To round off this section on slugs gardeners may be interested to hear that there are three species of predatory slugs in the British Isles. They eat earthworms and other soil invertebrates, including other slugs. They can be recognized firstly by a small external ear-shaped shell carried on their tail end, and secondly by their shape. These slugs taper in a most unslug-like way — from tail to head rather than head to tail as most slugs do. *Testacella haliotidea,* probably the most common shield-shelled slug, can be up to 10cm (4 in.) long and is either cream or pale brown. Because they are now rather uncommon they should not be killed.

Snails

Unlike slugs, snails possess a shell into which they can withdraw completely. They are less important pests than slugs but still troublesome in the garden. The conditions that favour slugs are also ideal for snails and like slugs they feed on plant material and are mainly nocturnal. They tend to be restricted to calcareous soils.

The same control methods outlined for the control of slugs are appropriate — with one difference. Snails are actively sought out by the song-thrush which breaks open the shell to reach the soft tissues inside. It does this by using a stone as an anvil, so strategically placed stones in the garden will act as a focal point for song-thrushes and will be used to break open snail shells.

Aphids

Aphids are small, sap-feeding insects which vary in colour from white through various shades of red, yellow, orange and green to brown, blue and black. They are often known as greenfly, blackfly or blight and should not be confused with whitefly although they feed in a similar way, spread plant viruses and fungi and are detrimental to plants. Damage to plants by aphids is a result of the colonies feeding on the young tissues, which weakens and distorts new growth, while the fouling of the leaves and stems with honeydew encourages the growth of sooty moulds and the transmission of viruses from diseased plants to healthy plants.

Aphid colonies usually consist of both winged and wingless

Fig. 23 Aphid

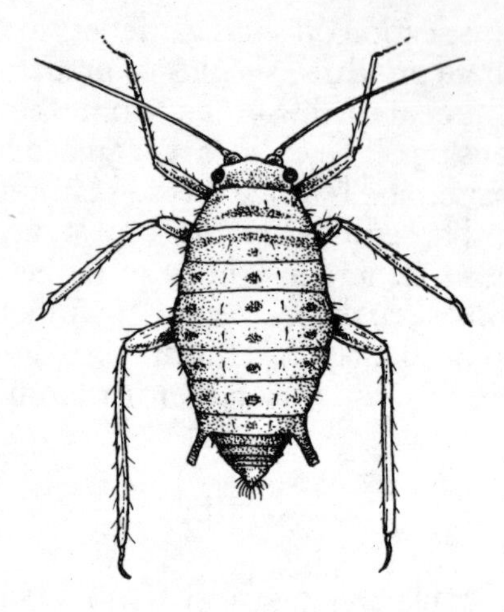

individuals. They are weak fliers but can be carried on thermals and air currents, which is how many species disperse and migrate to new plants. Both young and adult aphids feed on the sap of plants and colonies are commonly found on leaves, buds, stems, roots and flowers. Large volumes of sap are ingested and excess sugars and water are excreted as honeydew. Aphids are often tended by ants which feed on honeydew and give aphid colonies some protection from predators and parasites. Ants will drive away predators such as adult ladybirds and their larvae.

Aphid reproduction is mainly asexual (without a male for fertilization) and females are capable of giving birth to live young (without an egg stage as in other insects). As summer temperatures rise aphids mature in about a week which leads to a rapid build-up of aphid populations.

The same species can be an important pest of a number of plants. The peach-potato aphid, for instance, attacks not only peaches and potatoes but also lettuce, tomatoes, chrysanthemums, roses and many other plants.

Yellow decoy traps

Certain colours attract or repel aphids and this response can be used to control flying aphids. Aphids are repelled by the colour white because its reflective properties blur the insect's view of its target, whereas yellow attracts aphids because they associate the colour with young foliage.

Commercially available sticky yellow traps are coated with a non-drying glue and once the aphids land they are caught in the glue. Best results are obtained when the traps are hung near to plants and used at fairly high densities (one trap every 2–5 sq.m/25–50 sq.ft). They can also be used at a lower density for monitoring aphid populations. There are no insecticides in the trap and beneficial insects are not affected, so they can be used both indoors and outside.

Natural insecticides

In order of preference: derris; pyrethrum; quassia; nicotine — for use by professional growers only.

Spraying with insecticidal soap is effective in the control of aphids.

Microbial aphicides

Microbial aphidicides are now being used to control aphids on greenhouse crops. They contain the naturally occurring fungus *Verticillium lecanii*, which will attack only a restricted range of insects but is particularly effective against aphids and has a minimal effect, if any, on beneficial insects.

Verticillium lecanii has a requirement for high humidity, (about 85 per cent Relative Humidity and minimum night temperatures of 15°C (59°F)) and is therefore not recommended, at present, for outdoor use in the UK or in environments of low humidity.

It is harmless to livestock, wildlife and no allergic or other harmful effects on humans have been observed. *Verticillium lecanii* is not thought to be toxic to plants, although it is recommended that a small area of plants be tested first.

Verticillium lecanii should be applied early in the life of the crop when there are few aphids present. Those affected by the fungus will die in 10–14 days and the fungus will appear as white fluffy bodies, which will reinfect aphids not hit by the original spray. It is the fungal spores on dead aphids that give it its residual effect which may last for several months.

Alarm pheromones

Not all insect pheromones serve to attract individuals of the same species. Pheromones can also evoke alarm responses in insects and several alarm pheromones have been isolated from aphids.

Aphids which are disturbed release a pheromone which

stimulates other aphids in the vicinity to cease feeding and drop from the plant onto the ground or onto lower leaves.

Control of aphids using a contact insecticide is often difficult because the insects feed on the underside of leaves and it is difficult to direct spray droplets at them. However, when the alarm pheromone is mixed with the insecticide, aphids fall from their feeding sites on the underside of leaves onto the upper surface of lower leaves where they make contact with the insecticide.

Although alarm pheromones are not yet commercially available, a number of companies are investigating their potential.

Natural enemies

The natural enemies of aphids are numerous and can be divided into predators, parasites (parasitoids) and micro-organisms.

Predators are the most familiar group of the aphids' natural enemies. Examples include ladybirds, lacewings, parasitic wasps and hoverflies. While some are species-specific, most predators are able to feed on a large diversity of aphid species.

The larvae of a large number of hoverflies feed on aphids and make a considerable impact on aphid populations. Hover-fly larvae are voracious feeders and will dispose of many hundreds of aphids during the larval stages.

The smell of aphids is the most important stimulus which causes the female hoverfly to alight and lay eggs. Hoverflies seem to be less haphazard than ladybirds when laying their eggs, placing them closer to aphid colonies, so that hatching larvae have a plentiful supply of food.

Since the colour yellow is attractive to hoverflies as well as aphids, and hoverflies feed on a diet of pollen as well as nectar, yellow flowers such as broom can be planted near to crops.

Ladybird larvae are generally more fastidious than hoverflies regarding aphids. This is because some aphid species are poisonous to some ladybird species. During the larval stage, ladybirds dispose of many hundreds of aphids, and the adults also feed on aphids.

Some species of rove beetle and ground beetle are probably also important predators of aphids.

Both larvae and adult lacewings feed on aphids and con-sume many hundreds during their lifetime. The powdery lacewings, which are much smaller than the green and brown lacewings, prey on small aphids, as well as on scale insects and mites.

Some Braconids parasitize aphids — a single larva develops inside the host's body. They tend to be species-specific regarding prey, although not in all cases; for example, the cabbage aphid parasitic wasp will attack the peach-potato aphid but not the black bean aphid.

Many aphids overwinter usually as eggs and populations can be reduced by encouraging birds of the tit family into the garden. By hanging a small piece of fat amongst the roses or fruit trees, birds that are waiting their turn to feed on the fat will be searching amongst the bark for aphids and their eggs.

Intercropping

A number of successful intercropping trials have been conducted which have resulted in reductions of aphids.

Underplanting Brussels sprouts, cabbages and cauliflowers with red and white clover seems to reduce populations of aphids.

It appears that clean-weeding of brassicas provides ideal conditions for colonization by aphids. Brassicas planted in bare soil stand out against their background.

Interplanting crops with red or white clover visually masks the crop and also provides cover for predators. Intercropping cabbages with French marigolds (*Tagetes* spp.), however, may reduce cabbage yields.

Ant control

Ants protect aphids from their natural enemies to ensure that aphids continue to excrete honeydew which is attractive to the ants. Ladybirds and lacewing larvae are both driven off by ants, but hoverfly larvae protect themselves with a slimy exudate and are not actually driven off the plants. Similarly, some species of parasitic wasp are not attacked or disturbed by ants while laying their eggs. However, if ants are crawling on aphid-infested plants, you can assume that the ants are playing a protective role.

A barrier of non-toxic, non-drying sticky material (as used in moth traps, see page 46) can be applied to trees and the like, which will prevent ants passing. Alternatively, derris is effective against ants.

Other bug pests

There are a number of other types of bugs — lace bugs, capsid bugs, frog-hoppers, leaf-hoppers, scale insects and whiteflies. Not all are harmful; some species of capsid and anthocorid bugs prey on mites, aphids, caterpillars and other invertebrates. They are all piercing insects: both nymphs and adults pierce plant or animal tissues with fine stylets, inject saliva and ingest the resulting fluid.

Frog-hoppers

Frog-hoppers, as the name suggests, are rather frog-like both in appearance and habit. They have prominent eyes and powerful hind legs with which they jump when disturbed. The nymphs feed on plants under conspicuous coverings of froth, commonly known as 'cuckoo spit'.

The froth and nymphs can be removed from plants by spraying forcibly with water from a garden hose. Both adults and nymphs can be controlled with nicotine.

Capsid bugs

Some capsid bugs are beneficial and prey upon a number of invertebrates; others, however, damage leaves, buds, stems and developing fruit of a variety of plants including apple, pear, plum, currant, gooseberry, raspberry, strawberry, potato, runner bean, and ornamentals such as aster, chrysanthemum, dahlia, forsythia, hydrangea and many more.

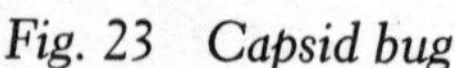

Fig. 23 Capsid bug

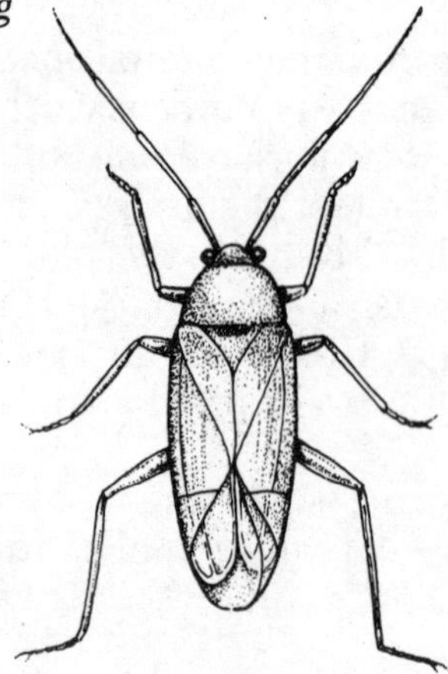

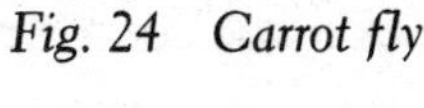

Fig. 24 Carrot fly

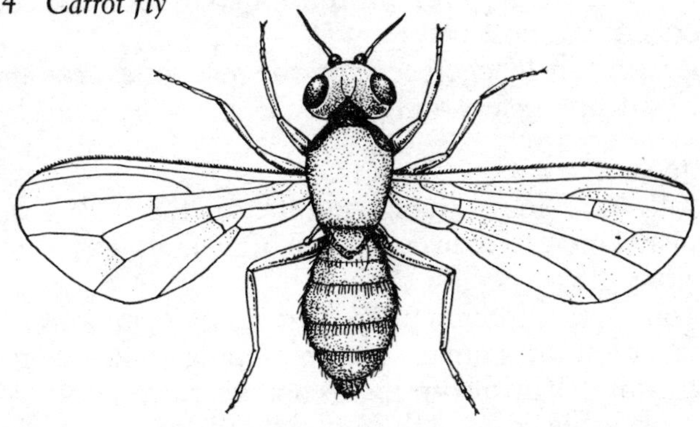

Capsid bugs are best controlled by spraying with nicotine (now available only to professional growers) in spring, summer and autumn when the pests are active. Overwintering eggs on deciduous apple trees, ornamental shrubs and the like may be killed with a winter wash in December or January while the plants are fully dormant.

Leaf-hoppers

Leaf-hoppers are similar to frog-hoppers and are responsible for spreading a number of plant diseases, such as green petal disease in strawberries. They attack various ornamental plants, fruit trees and bushes, vegetables such as potatoes and hornbeam and beech hedges. They are best controlled by spraying with nicotine.

Carrot fly

The carrot fly is a small (8mm long) two-winged fly with a red head and shining black body. The adult flies emerge first during late April and their numbers reach a peak in early June. There is a second generation of fly activity in most parts of Britain, beginning in August and continuing through to September. In some parts of Britain there may be a third generation beginning in late October.

The adult flies feed on nectar and after mating the female flies search out a host and lay their eggs in the soil close to the plant. Common wild hosts include hedge parsley (*Torilis japonica*), hemlock (*Conium maculatum*), fool's parsley (*Aethusa Cynapium*) and rough chervil (*Chaerophyllum temu-*

lentum). The larvae crawl down through the soil and feed off the roots of the host plant.

The carrot fly maggots feed on the roots of carrots and also attack parsnips, celery and parsley.

Intercropping

Carrot fly is not found in high numbers in the wild because its host plants grow in a diverse range of habitats surrounded by numerous other species.

In some years intercropping carrots with onions can help reduce carrot fly damage. Intercropping probably reduces carrot fly attack by disturbing its ability to detect host plants. This is probably because the onion volatiles mask those of the carrot, particularly when the onions are young. Intercropping, again particularly when the onions are young, increases the number of ground beetles and rove beetles which are predators of carrot fly eggs.

Intercropping with French marigold, *Tagetes patula* is ineffective.

Mulches

Mulching with lawn clippings sometimes helps to reduce carrot fly damage. This method probably works by confusing the fly's sense of smell. Yields may be reduced, however.

Barriers

Perhaps the best method of protecting carrots from carrot fly is to erect a barrier 80cm (32 in.) high around but not over the crop. Most adult carrot fly entering carrot plots are female and the majority approach upwind at a height of approximately 40cm (16 in.) above the ground, possibly in response to carrot odour.

The barrier frame should be 1–1.3m (3–4 ft) square and as sturdy as possible, with the sides covered in fine mesh shading, such as Papronet R, to exclude the flies. Alternatively, doubled-over polythene can be used. (See Fig. 26.)

It is also possible to avoid carrot fly attack by using plastic tunnels or cloches. Plants must be at least 50cm (20 in.) away from the wall of the cloche or tunnel because the larvae can migrate up to 50cm (20 in.) to reach a host plant.

Resistant varieties

Unfortunately at present no varieties of carrot exist which are

Fig. 25 Carrot fly barrier

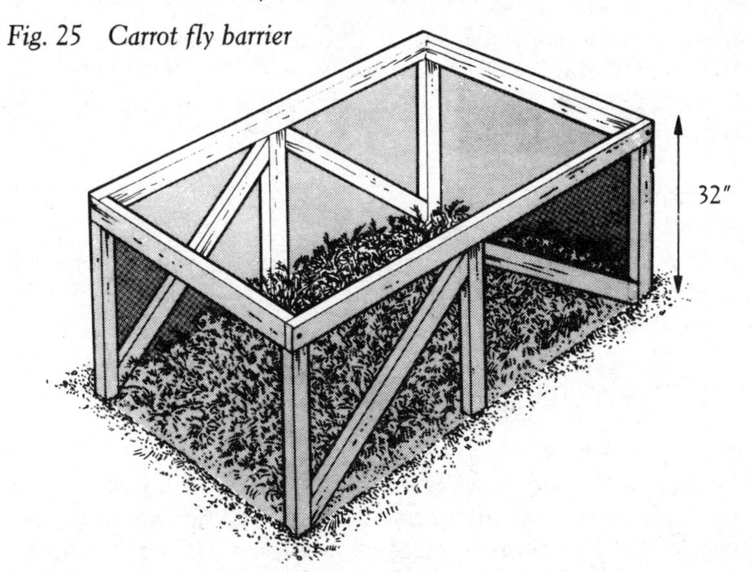

completely resistant to carrot fly. However, a few cultivars have proved to be partially resistant (see Appendix II) and may help to alleviate the carrot fly problem if combined with other control methods. It appears that resistant varieties release substantially fewer volatiles than susceptible carrots.

Natural insecticide
Derris.

Cultural control
Damage by the first generation of carrot fly can be reduced by delaying sowing, but yields may suffer because of the shorter season. Alternatively, early sowing avoids second generation damage and allows for early harvesting. This is important because damage to carrots increases over a period of time.

In southern Britain carrots sown in mid-March or mid-June will avoid the main egg-laying period of the first generation. In the north of Britain this generation is delayed by 10–14 days. So by harvesting early crops before the end of August and maincrops in October damage will be reduced.

Winter digging will expose some of the carrot fly maggots and pupae to birds and other predators and this reduces pest numbers.

The practice of thinning carrots should be avoided. The

damage to seedlings caused by thinning releases small quantities of attractants which carrot flies will quickly home in on. Crevices left in the soil after thinning also provide sites in which the female carrot fly can lay its eggs.

Monitoring traps

Carrot fly has two generations each year — first during May–June and then August–September — separated by a period of about a month when there is no fly activity. By using sticky traps (see page 75) it is possible to monitor the degree of infestation of carrot fly so that a spraying regime can be implemented.

Natural enemies

Predation of carrot fly eggs usually occurs in spring when small rove beetles are abundant and in autumn when small ground beetles abound. Some species of parasitic wasp and rove beetle attack carrot fly by parasitizing the larvae.

Cabbage root fly

The cabbage root fly is one of the most serious pests of brassica crops, and has two and sometimes three generations in one year from April to October. It lays its eggs in and on the soil around the base of the stems of brassica plants. The larvae feed on the roots, causing damage which may result in stunted growth and sometimes death of the plant.

Fig. 26 Cabbage root fly

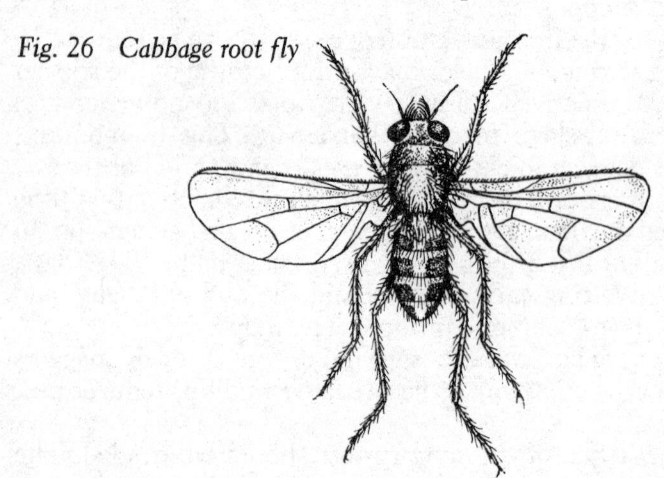

Each female fly may lay as many as 80 eggs and where flies are numerous hundreds of eggs may be laid around each plant, so there is a considerable potential for root damage. Fortunately, the mortality rate is very high, due to desiccation of the eggs in dry soil, severe root damage leading to the death of the host plant and consequently the larvae, and predation.

Before the mid 1970s cabbage root fly was relatively easy to control. However, since then vast areas of oil-seed rape have come under cultivation and a huge reservoir of root fly has built up.

Natural enemies

Several species of ground beetle and rove beetle are important predators of the immature stages of the cabbage root fly, but of these beetles some are more important simply because of their greater numbers. The most active beetles have the highest predator value, probably because eggs are largely found by contact. The small (active) *Bembidion* and *Trechus* species are probably the most important predators and will consume many hundreds of the eggs and larvae. Beetle numbers increase with increasing numbers of root fly eggs and decline as the number of eggs decline.

The larvae as well as the adults of rove beetles are predators which attack the eggs, larvae and pupal stage of the cabbage root fly. Rove beetles are also often more numerous than ground beetles.

The gall-wasp *Idiomorpha rapae* is a parasite of the cabbage root fly, and lays its eggs on the first or second larval stage, the adults emerging from the puparium.

Natural insecticide

Cabbage root fly is now resistant to insecticides in many areas and other control methods are recommended.

Cultural control

To reduce the numbers of cabbage root fly pupae overwintering in the soil, you should turn over the soil during the winter months to expose the pupae to predators such as the robin and other birds.

Another way to reduce root fly damage is to transplant brassicas early in the year so that plants are well established by the time the fly attacks. As the attack by the first generation of root fly is unlikely to be severe before May, crops transplanted in early February are usually large enough to withstand that attack. Feeding and regular watering also help to minimize the effects of crop damage.

The cabbage root fly is very catholic in its tastes and can develop on the roots of a wide range of cruciferous plants; for example, it thrives well on stocks and on pennycress.

Reducing the number of cruciferous plants may help, although this is probably impractical.

Intercropping

A number of successful intercropping trials have been conducted which have resulted in a reduction of damage to brassicas by the cabbage root fly.

Brussels sprouts underplanted with white clover visually masks the crop from the root fly, whilst intercropping cabbages with clovers provides cover for predators.

Single row intercropping appears to be the best arrangement of plants for reducing root fly attack and is most effective when the intercrop provides at least 50 per cent ground cover between the rows at the time of pest invasion.

It also appears that a mulch of green paper (or similar) also diminishes egg numbers.

Barriers

Adequate control of the cabbage root fly can be achieved by placing protective discs on the soil around the base of brassicas. The most effective discs are approximately 13cm (5 in.) in diameter cut from the foam-rubber type of carpet underlay. For the discs to be effective they have to fit closely around the plant stem and expand as the plant grows. (See Fig. 27.) They must also be applied directly after planting.

They appear to work for three reasons: firstly, they act as a physical barrier preventing the fly laying its eggs in the soil around the stems; secondly, the discs act as a mulch, conserving moisture around the roots of the plant and allowing the plants to withstand greater amounts of root damage; and thirdly, the humid microhabitat created beneath the discs encourages predatory beetles, such as ground beetles, to congregate. Consequently, these beetles eat proportionally more root fly eggs and larvae around plants protected by discs.

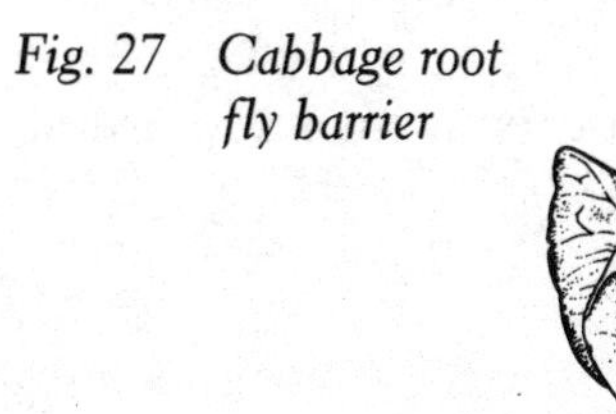

Fig. 27 Cabbage root fly barrier

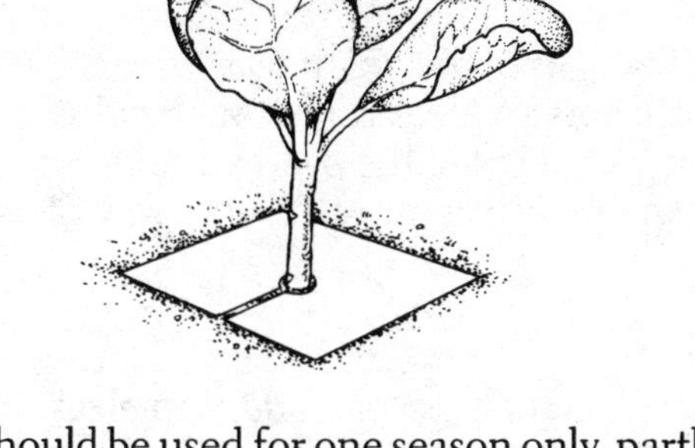

The discs should be used for one season only, partly because they become misshapen and stretched during the growth of the crop and also because they will deteriorate in time and contaminate the ground.

Planting brassicas through black polythene also works, but is not as effective as discs.

Other fly pests

Other fly pests include frit fly, celery fly, onion fly and crane-fly (daddy-long-legs or leather-jackets).

Frit fly

This is generally not a pest of gardens but does attack sweetcorn.

The larvae bore into the young seedlings in early May and cause tillering, distorted growth and reduced yields.

Crops sown or transplanted in late May are not usually affected. Good growing conditions should be provided so that plants can become established rapidly.

Frit flies also have a number of natural enemies, which include anthocorid bugs (*Anthocoris nemorum* and *Anthocoris nemoralis*), predatory flies (*Tachydromia minuta* and *Trachydromia agilis*) and predatory mites (*Pergamasus longicornis*).

Celery fly

This pest is also known as the celery leaf miner because the larvae tunnel within the leaves of celery and parsnips. Because of this habit the celery fly is difficult to control once established. Light attacks, however, can be checked by crushing larvae and pupae within the mines or removing and burning affected leaves.

Onion fly

The white maggots of the onion fly feed in the stems and bulbs of onions, shallots and leeks. Thorough cultivation during the winter months to expose the overwintering pupae to the birds and the elements is the most reliable way of controlling this pest.

Crane-fly

The larvae of crane-flies or daddy-long-legs, also known as leather-jackets, are particularly prevalent in newly cultivated grassland and wasteland and in moisture-retaining soils.

It may be necessary to devote the first season to clearing the ground before planting. A cleaning crop such as potatoes should be grown during the first year. The soil should be adequately drained because a moist, badly drained soil will attract flies and provide conditions favourable to egg-laying. All weeds should be destroyed and the soil disturbed by hoeing regularly which helps reduce the number of larvae. Grass turves placed grass-downwards collect larvae beneath them which may then be captured and destroyed.

Fig. 28 Crane-fly

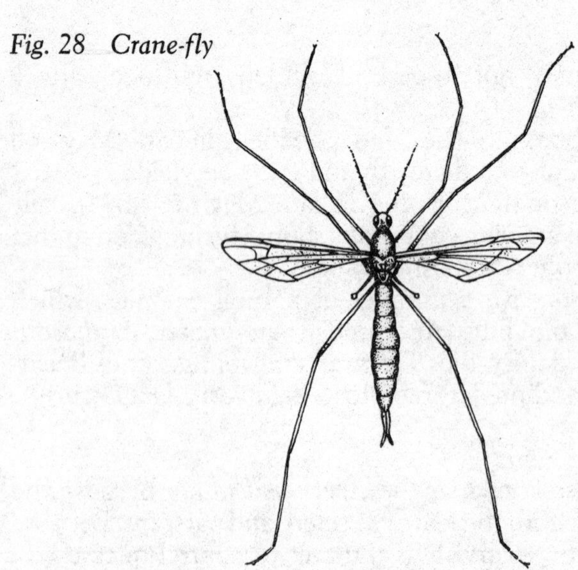

Fig. 29 Sawfly

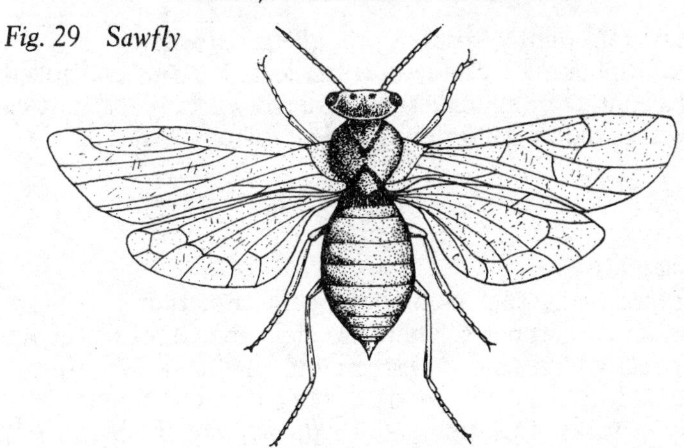

Sawflies

Adult sawflies feed mainly on pollen, but the larvae eat the leaves and developing fruitlets of various fruit trees and bushes and the leaves of ornamental plants.

The slug-like larvae are known as slug-worms.

Fruit-tree sawflies

Sawflies attack a number of fruit-bearing trees and bushes including apples, plums, damsons, pears, cherries, almonds, gooseberries and currants.

Sawfly larvae which attack apples, plums and damsons eat into the developing fruitlets. The trees should be sprayed with nicotine (available to professional growers only) thoroughly about a week after the petals have fallen. Timing is critical as the larvae must be killed before they tunnel into the developing fruitlets. Defer spraying until late evening as nicotine is lethal to bees.

Sawfly larvae which attack pears, cherries, almonds, gooseberries and currants as well as apples, feed on the leaves throughout the season. If caterpillars are seen spray them with derris or quassia, paying particular attention to the centre of bushes.

Ornamental sawflies

Sawflies attack a number of ornamental trees, bushes and plants, including hawthorn, rowan, birch, acer, willow, pine, hornbeam, hazel, rose, Solomon's-seal and iris.

They generally feed on the leaves and shoots (pine), but also roll leaves up tightly (rose) or produce galls (willow).

Those species which feed off the leaves can be controlled using a contact insecticide such as derris. Pick off and destroy leaves affected by gall-forming or leaf-rolling sawflies.

Beetles

Flea beetles

There are many species of flea beetle, but only a few are important garden pests. There are eight species of the genus *Phyllotreta* which feed on cruciferous plants such as turnip, swede, radish and cabbage, six of which are common.

There are also two species of *Psylliodes*, one of which feeds on cabbages and the other on potatoes.

The adults of *Phyllotreta* hibernate from October to March in hedgerows and other available shelter. As the temperature rises in spring they move out of hibernation sites into adjoining land. They are active on fine days towards the end of April, after which they disperse and become widespread. They feed on the cotyledons and slender stems, attacking them while the plants are still underground, so that the plants wither.

Towards the end of May the beetles become less active, pair and lay eggs usually in the soil. The larvae eat the roots or stems and pupate in about four weeks. Adults emerge at the end of July and early August when they feed on the mature crop.

Some species transmit radish mosaic and turnip crinkle viruses.

Fig. 30 Flea beetle

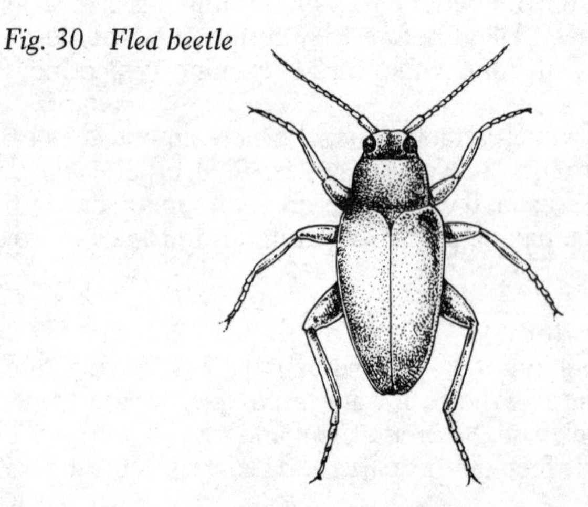

Cultural control
Most damage is done to seedlings during April and May. In moist weather, plants grow better and escape serious injury, but in dry conditions damaged plants will wilt and may die. Water and feed to encourage rapid growth and prevent desiccation. Crops sown early (March) or late (June) will avoid the peak infestation periods.

Natural insecticides
Derris; pyrethrum.

Natural enemies
Natural enemies of the flea beetles include the ichneumon and Braconid parasitic wasps. (In the USA predatory nematodes are used to control flea beetles.)

Sticky boards
Card coated with a non-drying sticky adhesive can be hung or carried close to ground level between rows of infested crops. These work because of the beetles' response to disturbance: when flea beetles are disturbed they respond by jumping away from the site of danger.

By walking between the rows of crops and gently brushing each plant the beetles will respond by jumping into the air and getting caught on the adhesive cards.

Intercropping
Intercropping trials have been conducted in the USA on the control of the flea beetle *Phyllotreta cruciferae*. The results may not be valid in UK conditions, but are probably worth trying.

Intercropping brassicas with wild cruciferous plants (*Erysimum chieranthoides*, and *Iberis amara*) may reduce populations of flea beetles.

Barriers
Plants covered in Agryl P17, a lightweight polypropylene cloth, may help reduce populations of flea beetles.

Strawberry beetles
There are two types of strawberry beetle: those which remove the seeds and those which bite holes in the flesh. All are ground beetles.

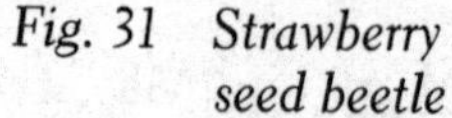

Fig. 31 Strawberry seed beetle

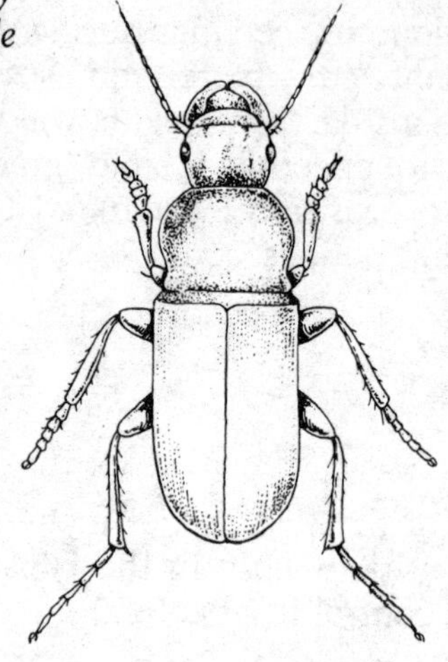

The adults and larvae of *Harpalus rufipes* will eat the seeds of strawberries. The adults of *Pterostichus madidus* and *Pterostichus melanarius* eat the flesh; the larvae are exclusively carnivorous.

The extent of damage by *Pterostichus madidus* is determined not only by hunger and thirst, but also factors which affect the beetles' surface activity. The strawberry ground beetles are more active after rain and *Pterostichus madidus* inactive during periods of drought. A dry period will result in the beetles becoming both starved and desiccated. When it rains, therefore, they will feed readily, so that attack is likely to be most severe at this time.

The preferred food of the strawberry seed beetle (*Harpalus rufipes*) is the seeds and plant material of fat hen, duckweed and other weed plants. *Harpalus rufipes* is also an important predator of small white butterfly caterpillars and is known to eat slugs. The strawberry beetles *Pterostichus madidus* and *Pterostichus melanarius* are rather catholic in their tastes and will include slugs and other arthropods in their diet.

As these ground beetles are also three of the most common and widely ranging species, it may be best to consider them as 'beneficial' and only try to control them actively where strawberries are cultivated.

Natural enemies
Nocturnal insectivores such as hedgehogs, shrews, and owls.

Traps
A canister without a lid and preferably without a lip — such as a large baked bean tin with the lid cleanly removed — is sunk into the ground so that its mouth is level with the soil. To prevent rusting such tins can be painted with Hammerite or similar paint.

The traps should be inspected daily. The beetles should be removed to a site a long way from the strawberry patch or emptied into the slug trap (see page 58).

Barriers
A barrier similar to that recommended for carrot fly control (see page 67) may also help to reduce the number of strawberry beetles. The barrier should be approximately 30cm (12 in.) high and surround the strawberry patch. Pitfall traps inside the barrier will help to catch any remaining beetles.

Wireworms
Wireworms are the larvae of various species of click beetle. These beetles are easily recognized because of their ability to flick themselves into the air when they need to right themselves. They can take up to five years to reach maturity, during which time they feed on the roots, corms, tubers and stems of many plants.

The larvae develop mainly in grassland but are occasionally troublesome in gardens in close proximity to undisturbed land

Fig. 32 Click beetle and wireworm

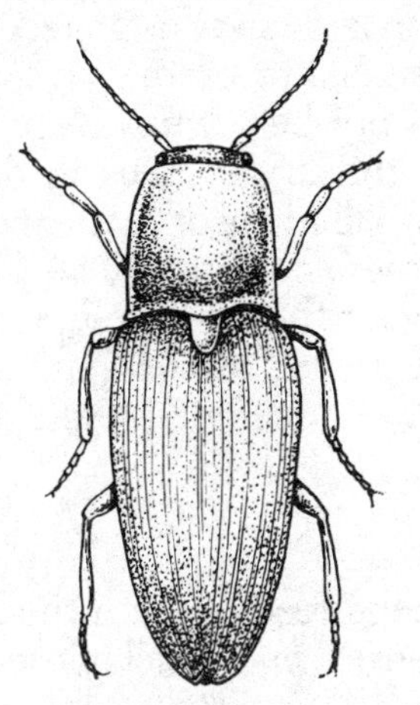

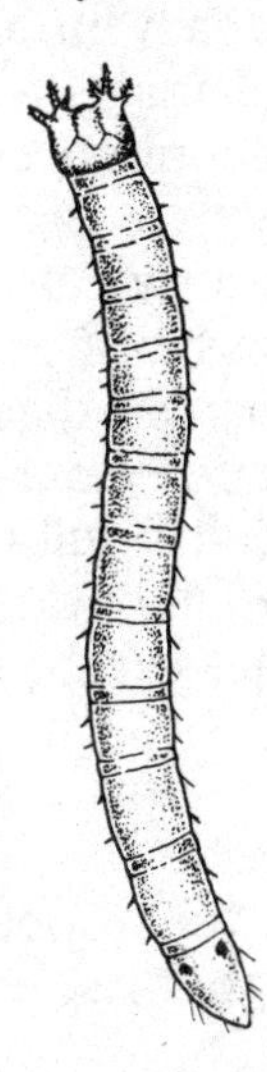

such as orchards or wasteland, or in newly cultivated ground. Thorough cultivation and good weed control should eliminate wireworms from newly cultivated ground. Avoid growing potatoes and other susceptible crops; instead grow crops such as parsnips, spinach or beans which are less attractive to wireworms. If you must plant potatoes lift them early, since damage will be greatest when they are left too long in the soil. Alternatively, remove the top 7cm (3 in.) of turf in early spring as the wireworms are closest to the surface at this time of the year. As the turf is removed it should be stacked in layers, grass side downwards, checked periodically and exposed to the birds — the wireworms' natural enemy. If turf heaps are allowed to get covered with grass, wireworms will only increase, so cover heaps with black polythene.

Other beetle pests

Other beetle pests include chafer beetles, weevils, raspberry beetle, asparagus beetle.

Chafer beetles

Cockchafers, also known as 'May bugs' or 'June bugs' feed on the foliage and flowers of various plants and do little damage; the larvae, however, damage the roots of turf or arable crops. The larvae are easily recognized, as they are broad and fleshy and curved like the letter C with the last abdominal segment enlarged.

Before the advent of synthetic pesticides they caused considerable damage to crops, particularly in newly cultivated grassland, the adults occurring in large swarms. In recent years, however, they have not been such a problem.

Gardens near woods should not be sown with potatoes if grubs are found in the soil. As the larvae are easily damaged, use of a mechanical digger may kill many of them and expose grubs to the birds. Rooks, magpies and thrushes all prey upon the larvae. Poultry will also devour the larvae and, if possible, should be turned out onto the ground after it has been dug.

Adult beetles can be killed while feeding by spraying with derris.

Raspberry beetle

The larvae of the raspberry beetle feed in or on the ripening fruits of raspberries, loganberries and blackberries. Adult raspberry beetles may also cause damage by feeding on buds and flowers earlier in the year. Beetles seen on the plants during May or June should be controlled with derris.

Asparagus beetle

Adult beetles and their larvae feed on the shoots and foliage of asparagus plants. Populations of overwintering beetles can be reduced by clearing away plant debris thus reducing hibernation sites. Adult beetles and their larvae can be controlled with derris.

Weevils

Weevils are beetles which have a characteristic snout projecting forward from the head, on the end of which are the mouthparts. There are many species that occur in gardens, most of which are minor pests. The most notable pest species are the apple-blossom weevil, turnip-gall weevil, pea and bean weevil and the glasshouse vine weevil.

The pea and bean weevils eat small scalloped pieces out of the leaves of both peas and beans in spring and summer which can seriously damage seedlings — older plants do not suffer much damage. Protect young plants by dusting the leaves with derris or trap weevils along pea or bean rows by using pitfall traps (see page 58).

The turnip-gall weevil is much more difficult to control as the larvae develop in the roots of brassica crops, forming galls. It is best to remove and destroy the roots of old brassicas in early winter and cultivate the soil, exposing the pupae to the birds.

The apple-blossom weevil does not normally cause appreciable damage as it reduces the amount of fruit setting and therefore acts as a natural thinning agent. The larvae develop in each flower and after feeding for a few weeks pupate under the dead petals. Only when the adults emerge in June or July and feed on the leaves can they be controlled, using a non-persistent insecticide.

Another weevil pest is the vine weevil (see page 92).

Woodlice

Woodlice (or slaters or pillbugs, as they are also known) occur in large numbers in gardens and especially in greenhouses or

Fig. 33 Woodlouse

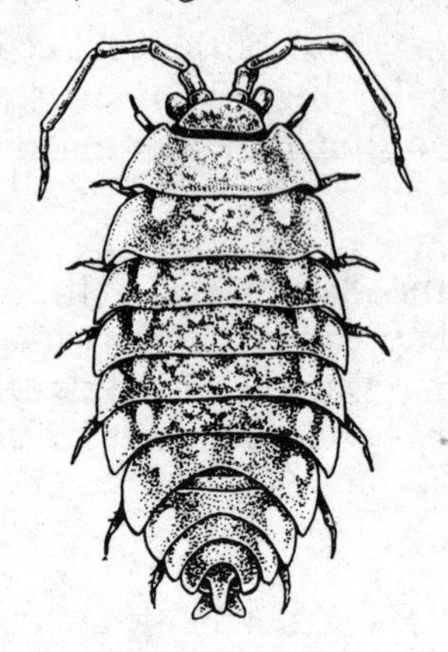

sometimes in conservatories. They are particularly abundant where it is damp and shady and where there is much decaying organic matter.

The presence of decomposing garden rubbish, decaying wood or the like provide conditions favourable to woodlice. They are nocturnal creatures and hide away during the daytime in crevices or anywhere damp and dark.

Their food consists chiefly of decaying vegetable matter, but when the supply is exhausted they attack living plants and gnaw at the stems, devour roots and eat out irregular holes in the foliage of a great variety of garden plants in the open and under glass. Their presence may well be unsuspected because of their nocturnal activities and injury to plants is often thought to be due to some other pest.

Cultural control

Cleanliness is perhaps the best method of control. Clean away and burn any debris, especially in greenhouses. Boards, bricks and flowerpots left around will just provide shelter for woodlice during the daytime. In stone-floored conservatories woodlice sometimes become a pest, occurring in large numbers; the vacuum cleaner is perhaps the best tool for extracting woodlice from crevices in the conservatory structure. Woodlice can also be reduced considerably by trapping them in scooped-out potato tuber halves or oranges under pieces of slate or board. These traps can be examined daily and the creatures removed.

Natural enemies

A frog or toad in the greenhouse will help to reduce woodlice populations. As regards other predators, a paradox exists in that a variety of animals are known to eat woodlice under laboratory conditions but there is little evidence of sustained predation in the wild. In captivity woodlice are eaten by shrews, toads, hedgehogs, slow-worms and frogs. Among the invertebrates which feed on woodlice in captivity are ground beetles, rove beetles, spiders, harvestmen and centipedes.

Earwigs

Earwigs sometimes do damage in orchards, particularly apple and pear orchards. They usually make small round holes in fruit which are subsequently enlarged by wasps or become infected by fungal growth, both of which lead to rotten fruit on the tree. In some cases brown rot fungus becomes established and spreads to nearby fruit through small russet cracks. Earwigs may also do damage in greenhouses by feeding on the foliage and blooms of chrysanthemums, dahlias and other ornamentals. They feed after dark, and hide away during the daytime in the leaves, petals and supporting bamboo-canes.

Large numbers of earwigs can be trapped in clay pots stuffed with straw and inverted on the ends of canes or rolled-up corrugated cardboard on the ground.

The Earwig Nest works in a similar way to the inverted clay pot and corrugated cardboard but is reputed to be more efficient. (See Appendix II for suppliers.)

As earwigs are also beneficial insects (see page 12), traps can

Fig. 34 Earwig

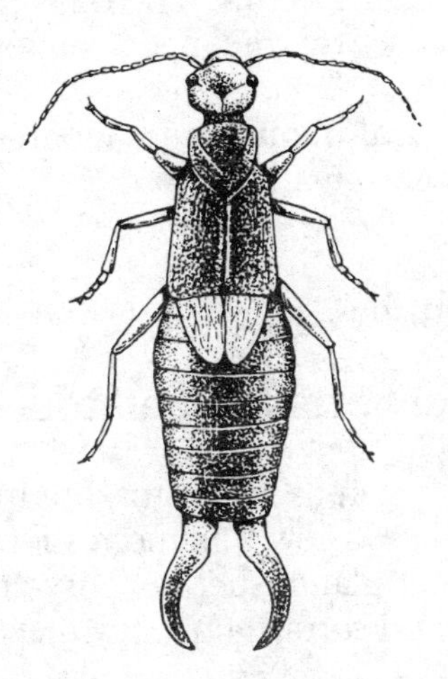

be emptied away from the greenhouse or other areas of infestation.

Millipedes

Millipedes are sluggish in their habits and, when disturbed, coil themselves up like a watchspring. They should not be confused with centipedes which are active and run away when disturbed.

The most injurious species are the rounded *Blanjulus guttulatus* and the flattened millipede *Polydesmus angustus*. These feed on a variety of vegetable matter both living and dead. They tunnel into seeds, attack seedlings, gnaw at the roots and underground stems of a great variety of plants and eat into bulbs, tubers and corms. They also enlarge the holes left by slugs, wireworms and other soil insects, penetrating the first borings and extending the initial injury.

Cultural control

Constant disturbance of the soil by hoeing and general cultivation will help to reduce populations of millipedes, because they dislike movement of the surface soil, which exposes them to predators.

Traps

Trapping by means of scooped-out potato tubers and carrots placed on a wooden skewer and buried beneath the soil surface is successful in smaller gardens. The traps must be examined every three days or so, the millipedes removed and the baits reset.

Garden refuse, leaf mould and manure which contain millipedes should be spread out to dry.

Natural enemies

Blackbirds and thrushes.

Eelworms (Nematodes)

Nematodes, also known as eelworms, are microscopic, translucent and worm-like, a few millimetres long and are mainly pests of ornamental plants such as chrysanthemums, bulbs such as daffodils, and shrubs, but they may also affect plants such as strawberries and blackcurrants. Severe infestations may kill plants.

Intercropping

Investigations have demonstrated that the cultivation of certain species of Compositae reduce or suppress populations of plant parasitic nematodes. Apparently certain compounds permeate out from the roots and into the nematodes. Nematodes up to 5cm (2 in.) away can be affected. However, it is not known how these compounds are activated, or their mode of action and unfortunately not all plant parasitic nematodes are suppressed — potato cyst eelworm, for instance.

One of the most effective plants found to suppress nematode activity is *Tagetes patula*. The wild variety is most effective — it seems that the more highly bred the plant the less well it works.

Tagetes, however, may reduce yields if used as an intercrop with plants such as cabbages. It is probably best to use it as a cleaning crop.

Helenium hybrid 'Moerheim Beauty', *Gaillardia* hybrid 'Burgunder', *Schkuhria senecioides* and *Ambrosia trifida* also suppress certain species of eelworms.

Cultural control

Eelworms are frequently introduced into gardens in plants already infected and in the soil adhering to them. If infected plants are moved, then infection may be spread. Infection can also be spread by people walking over infected ground and transporting earth on the soles of footwear from one part of the garden to another. Surface drainage water and the dirt on garden implements can also be a source of eelworm infection.

All weeds should be destroyed by cultivation to eradicate any possible wild host plants. All diseased plants should be burned and not thrown on the compost heap. No plant known to be attractive to a particular type of eelworm should be planted in infected soil for at least three years after an attack has occurred.

Cuttings from diseased plants are liable to be infected and it is best to raise new plants from seed where possible.

4
Greenhouse pests

At one time gardeners kept frogs and toads or encouraged spiders to live in the greenhouse to help control pests. Although this is still a valid control measure, biological control has become a great deal more sophisticated since then.

Biological control relies on using a pest's natural enemies to control it. These natural enemies may be parasites that kill it or stop it reproducing, or they may be predators that eat the pest.

An important point to remember is that most natural enemies are species-specific: in other words, they attack only one type of pest and usually only one species. So they will not harm the plants, wildlife or other insects, especially beneficial insects.

However, there are some drawbacks. This control is only really effective where the movements of both predator and pest can be controlled, such as in the confines of a greenhouse, and where temperature and to a lesser extent humidity can also be controlled. It can also work out quite expensive.

Biological control

Natural populations of animals fluctuate in size and are kept in check by predation, disease and the availability of food. Greenhouse pests increase rapidly in numbers during the summer when temperatures are high and food is plentiful. This natural increase may come to a halt when the food runs out, but more often is limited by predators and parasites which themselves increase when their food (the pest) is abundant. For example, an increase in aphids is often followed by an increase in ladybirds which feed on them and help limit their numbers.

Biological control of pests on perennial plants can be accomplished all year round as long as some pests are always present to support a population of natural enemies. Biological control on annual plants, however, usually means releasing new stocks of natural enemies every year.

The time of release is crucial, as biological control acts slowly. Natural enemies must be released when pest numbers are low, so that they have time to build up their own numbers in order to reduce the pests before they cause serious damage. The natural enemy will then also decline due to starvation. Low numbers of both pests and natural enemy should persist and further increases in pests will be prevented. If all the natural enemies die when the pest is eliminated then more must be introduced if fresh pests are noticed.

Pests and their control

Spider mites

The common red spider mite (*Tetranychus urticae*) gets its name from the brick-red colour of the overwintering stage which distinguishes it from the greenish-yellow summer active mites. The carmine spider mite (*Tetranychus cinnabarinus*) is dark red and does not have an overwintering stage. Spider mites attack many types of plants. They feed mostly on the undersides of leaves, sucking the sap so that the leaves appear mottled and eventually become brown and shrivelled. They also produce silken strands which cover the plants when infestation is heavy.

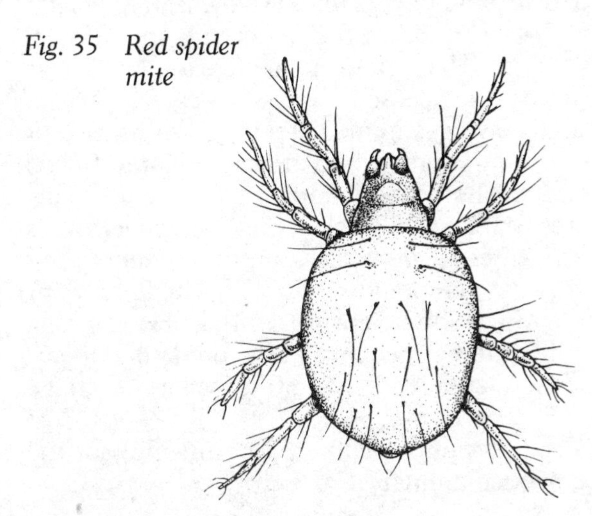

Fig. 35 Red spider mite

Their life cycle from egg to adult can take less than three weeks, so under certain — especially hot — conditions spider mites can multiply very rapidly, causing serious damage. However, a very humid atmosphere can slow this increase.

Control

The predator used to control the spider mite is itself a mite, the predatory mite *Phytoseiulus persimilis*. It is a fast moving, orange-brown, pear-shaped mite, slightly larger than its prey. It is very efficient at searching out its prey and eats both eggs and adults. It develops in half the time taken by the spider mites. The more predators in relation to the prey the sooner control is achieved. Once its prey has been eliminated, *Phytoseiulus* may survive for about three weeks to attack any fresh spider mite infestations. If these do not occur *Phytoseiulus* will die out and may need to be reintroduced.

To use *Phytoseiulus* to control spider mites, the undersides of leaves need to be examined at frequent intervals for the first signs of mite damage; a hand lens may be necessary for this operation. As soon as damage is observed, order *Phytoseiulus* immediately, stating the size of the greenhouse and size of infestation.

Phytoseiulus is supplied on bean leaves with a few spider mites as food, or in a handy shaker bottle. If supplied in leaf form, distribute one leaf on each infected plant. Predators supplied in containers can be uniformly shaken over the crop. It is important to distribute *Phytoseiulus* evenly, as they will not spread until they have eliminated the spider mite in one area, so any infestation missed will continue to increase. As soon as the predatory mites are released they will begin attacking spider mites. Adults lay eggs immediately and within two or three days the emerging young will go in search of prey. In two or three weeks there should be lots of predators and many fewer spider mites. Within 2–6 weeks of release, spider mites should have disappeared — depending on the initial ratio of predators to prey and temperature (25°C/77°F is best). A minimum of 13°C/56°F should ensure control all season. If a further spider mite outbreak occurs and the predator appears to be absent, a new supply of predators will need to be introduced.

As *Phytoseiulus* does not overwinter, a fresh supply of the predator will be needed annually.

Whitefly

The whitefly *Trialeurodes vaporariorum* is an important greenhouse pest of tomatoes and cucumbers, but also affects ornamental plants such as fuchsias. The adult whitefly can usually be found at the top of the plant while the larvae are found lower down. They are sap feeders and excrete vast quantities of honeydew on which sooty moulds grow. This makes the plants unsightly and affected foliage eventually becomes yellow or dies.

Adult whitefly are small (1mm long), winged insects covered in wax, which gives them a pure white appearance. Their eggs are laid mostly on the undersides of leaves and hatch into minute, scale-like nymphs which then disperse before feeding. These scale-like nymphs are at first mobile, but later become firmly anchored to the leaf where they pass through three larval stages and a pupal stage before completing their development.

Both adults and nymphs feed by inserting their stylets into the leaf tissues and sucking the sap. At 18°C/66°F the complete life cycle takes about 42 days.

Whitefly can survive freezing conditions for up to a month, so they may survive in the greenhouse during the winter and infect young plants in the spring.

Control

Because whitefly can withstand freezing temperatures, strict hygiene measures are necessary, particularly in the propagating house.

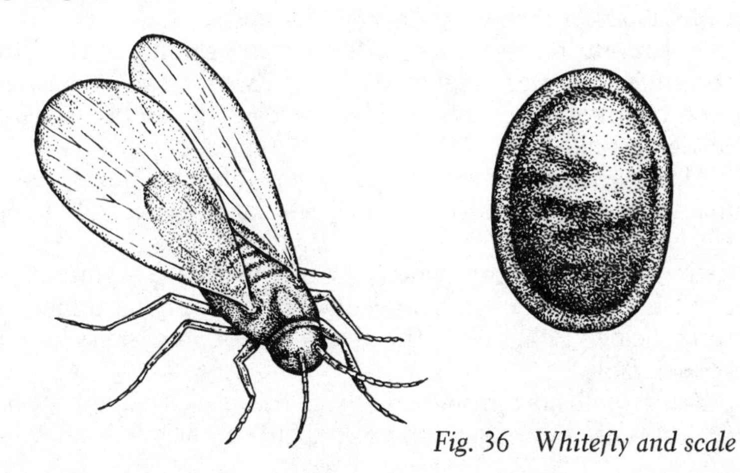

Fig. 36 Whitefly and scale

The whitefly's natural enemy is a parasitic wasp *Encarsia formosa*. These wasps have a remarkable ability for searching out and parasitizing the immature stages of whitefly, known as 'scale'. The female wasp reproduces parthenogenetically, in other words without male fertilization. Each egg is inserted into a different whitefly scale and the young parasite feeds on the larva within the scale. As the parasite develops, the scale turns black which confirms the successful establishment of the parasite. The adult parasite later emerges by cutting a hole in the roof of the scale.

Encarsia formosa is sold as black parasitized whitefly scales on fresh tobacco leaves or on small pieces of leaf stuck onto card. Hang the cards or cut-up leaves halfway up the infested plants away from direct sunlight and areas which are constantly wet. After about three weeks, check to see if the parasites have emerged by holding the card or leaf piece up to the light and observing the small exit holes in the scales. At least 75 per cent of the parasite should emerge.

The first new parasitized black scales should appear on the undersides of leaves in two to five weeks, depending on the temperature. For best results, the greenhouse temperature needs to be above 21°C/70°F, so that the parasite breeds faster than the whitefly. Under these conditions, 25 parasites per square metre (yard) of greenhouse floor area should control an evenly distributed whitefly population.

Encarsia formosa does not overwinter below 13°C/56°F, so new supplies of the parasite will be needed each year.

Limitations in the use of Encarsia formosa

The parasite is very susceptible to many chemicals used to control insects and fungi. If used with care the least toxic as a spot treatment are derris and pirimicarb (a synthetic manmade pesticide).

Heating systems using gas or paraffin without well-maintained flues can emit enough sulphur to kill adult *Encarsia*.

In severe infestations when a lot of 'honeydew' is produced, adult *Encarsia* are unable to work effectively and at temperatures below 15°C/59°F the development rate slows down considerably.

A naturally occurring microbial fungus *Cephalosporium leconii* is effective against whitefly and can be used safely in greenhouses where *Encarsia formosa, Phytoseiulus persimilis* or other commercially available biological control agent is in use. As well as being harmless to wildlife, it leaves no soil residue and is not phytotoxic.

Long-lasting pest control within the crop occurs because of the fungus's ability to spread. Whitefly not hit by the original spray are infected by the spores and hypae growing on the leaf surface and on the dead individuals.

Effective control can only be achieved with a continuous minimum temperature of 15°C/59°F, with humidity levels above 85 per cent Relative Humidity. The best control will be achieved if the underside of the plants' top leaves are sprayed. Whitefly affected by the fungus will die in about two weeks and will appear as white fluffy bodies.

Although *Cephalosporium leconii* is believed to be safe to all plants, it is best to test a small area before treating the whole crop.

Natural insecticides

For whitefly, use derris; for scale, use derris or nicotine (available to professional growers only).

Insecticidal soap can also be used.

Aphids

Use the microbial aphidicide *Verticillium lecanii* (see page 61).

Caterpillars (tomato moth)

Use the microbial insecticide *Bacillus thuringiensis* (see page 44).

Thrips

Thrips occasionally cause damage to foliage and fruit. They can, however, cause serious damage to peppers, especially around the calyx, which results in the growth of deformed fruit.

Thrips seldom fly except when disturbed and are usually found crawling over the leaf surface. The females are parthenogenetic and both the adults and larvae feed on the leaves, causing a silvery flecking on the upper surface and, in severe infestations, on the ripe fruit also.

Control

Amblyseuis cucumeris is a predatory mite which can keep thrip

Fig. 37 Thrip

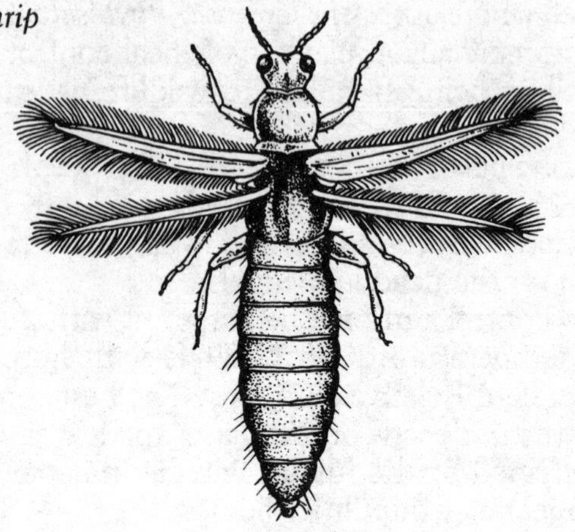

damage at a low level. It is pale pink in colour, similar to the predatory mite *Phytoseiulus* and with a comparable life cycle. The eggs are laid on the hairs in the axils and side veins underneath the leaf.

The predatory mite needs to be introduced into the crop at an early stage, so that there is a large population capable of immediately suppressing the thrips when they first appear. The predatory mite remains in the crop throughout the season, even when there are very few thrips.

Amblyseuis cucumeris does not tolerate chemicals used to control insects and fungi.

Natural insecticide
Use insecticidal soap.

Leaf miners
The tomato leaf miner (*Liriomyza bryoniae*) affects not only tomatoes, aubergines, peppers and beans, but also flowers such as chrysanthemum, gypsophilia and gerbera.

A second species, *Liriomyza trifolii,* has appeared in recent years and like the former species is able to destroy entire crops.

The flies are yellow-black in colour and about 2mm in length. The females lay their eggs in punctures made during the feeding process. The larvae eat tunnels just below the leaf surface known as 'mines'. These tunnels may occur in large numbers, reducing the green photosynthetic surface of the leaf. The fully grown larva leaves the tunnel and drops to the ground where it pupates. The whole cycle from egg to fly takes about three weeks.

Control

There are two indigenous hymenopteran parasites which can be used to control leaf miners. *Dacnusa sibirica* is an endoparasite which parasitizes by laying an egg in the leaf miner larva. The larva appears to develop normally and pupates in the soil. However, the parasite has been developing within the larva and instead of the leaf miner, a parasite emerges from the pupa.

Diglyphus isaea is an ectoparasite (wasp) which kills the leaf miner in its tunnel and lays its egg beside it. The wasp develops inside the tunnel while feeding on the dead larva.

It is difficult to determine to what extent parasites are active in the greenhouse and leaf samples need to be collected and analysed. Introduction of the parasites depends on the level of infestation.

Mealy-bugs

There are several species of mealy-bug, but the pest that concerns us here is the common or greenhouse mealy-bug, *Planococcus atri*. It attacks cacti, succulents, vines and ornamental plants including gardenias, amaryllis and orchids. The pests look unsightly and contaminate plants with honeydew on which sooty moulds grow. Mealy-bugs become established more quickly at high temperatures and in high humidities, and there may be up to eight generations each year.

Control

A predatory ladybird *Cryptolaemus montrouzieri* is used to control mealy-bugs. The black and orange adult beetle resembles a typical ladybird, but its larva is covered with a white wax and is similar in appearance to the mealy-bug itself, only much larger. Both adult and larva feed on mealy-bugs.

Cryptolaemus montrouzieri is not easy to rear and it is only supplied in small batches, enough to cover about ten plants. They are supplied either as larvae, pupae or adults. For best control, a temperature of 20–26°C/68–78°F and a high humidity is required. *Cryptolaemus* should reduce the pest population to a low level in two or three months, but may need to be reintroduced to maintain control.

Natural insecticide
Insecticidal soap can be used.

Black vine weevil (Otiorhynchus sulcatus)
The beetle feeds on the leaves of various plants, including begonia and cyclamen. It eats angular holes on the edges of the leaves. The larvae live near the roots of the plants where they cause far more damage. The larger larvae may bore into the main roots or tuber, or even eat away the bark around the stem base. This leads to wilting of the plant and its subsequent death. Although all stages of the life cycle can be found in the greenhouse at the same time, the adult beetles are generally encountered in the spring.

Control
A parasitic nematode is used to control the black vine weevil. It penetrates the larvae through the mouth, anus or respiratory openings. As it feeds, the nematode releases specific bacteria from its intestinal tract. These multiply rapidly and the beetle larva dies of blood poisoning and turns a characteristic brownish-red colour. When the nematode reaches maturity it multiplies and eventually the entire beetle larva is full of nematodes. These eventually leave the host in search of new beetle larvae. The survival of the nematode is largely dependent on the soil temperature and humidity. They do not function at less than 13°C/56°F.

Application is made with a conventional spraying equipment or irrigation system.

Rear or buy?

The convenience of having biological agents continually available must be contrasted with the likely cost of control by other means or with the purchase of natural enemies from commercial suppliers. Using these criteria home production is not attractive for small acreages, although a group of enthusiasts should carefully consider the possibility of collective action. Information on rearing parasites and predators can be obtained from the Glasshouse Crops Research Institute (see Appendix III). Suppliers of control agents are also listed in Appendix III.

5
Weeds

Weed is the name given to any plant growing where it is not wanted by man. A dandelion may be a weed in the lawn or vegetable plot but not necessarily in the wild flower garden.

Weeds can generally be divided into perennial weeds and annual weeds. Perennial weeds spread in several ways: some send out creeping stems above ground, for example creeping buttercup and slender speedwell; others produce bulbils, for example oxalis and lesser celandine; fleshy taproots which re-sprout are represented by dandelions and docks, while couch grass, bindweed, creeping thistle and ground elder are examples of weeds which have underground creeping roots or stems.

On the other hand annual weeds (such as shepherd's purse, and groundsel) generally spread by having wind-blown seeds or seeds which are dispersed explosively, for example petty spurge and hairy butter-cress. Of course, perennials such as dandelion are also spread by seed.

Most soils contain huge numbers of weed seeds. The soil in a neglected garden can easily contain thousands of seeds per square metre (yard). Seeds can remain dormant in the soil for years: on average only 2–10 per cent of the weeds in the soil will germinate in any one year.

Mulching

Mulching to irradicate weeds works on the principle of depriving the plant of light so that eventually the roots become exhausted and the weeds die. This does take a long time, at least one growing season, but it is perhaps the best way of *clearing* large areas of ground.

Mulching does have the disadvantage of encouraging slugs. Perhaps the best material for excluding light is heavy gauge, black horticultural polythene. It is supplied in rolls and can be re-used. It is best held down by burying the edges in the soil and weighing it down where necessary with sand bags filled with gravel. Alternatively old carpets, cardboard or newspaper can be used and they have the advantage of being both cheap and biodegradable. They do, however, have a number of disadvantages: firstly, they can look unsightly; secondly, they present ideal conditions for slugs, as the humidity is high and there is plenty of rotting vegetation for food; and thirdly, because cardboard and newspaper tend to be high in carbon their use could result in the soil being robbed of some of its nitrogen as they break down.

Old carpet must be made of material from natural fibres such as wool, as foam-backed carpets and carpets made from man-made fibres do not degrade and will eventually contaminate the ground.

Old carpet is one of the easiest materials to hold down in windy situations; a few pegs made of wire bent into hoops, or bricks, will hold the carpet in place.

Large cardboard boxes flattened out are very effective but the edges must overlap each other or weeds will push through the gaps. Holding them down can be a problem, although lots of bricks strategically placed should solve the problem.

Whole newspapers opened out and held down by a good covering of straw should suppress weeds for a while and these have the advantage of being flexible and small enough for mulching between shrubs, vegetables or fruit bushes.

However, do not use paper from glossy magazines as these contain a higher proportion of lead than newspapers and take longer to rot down.

The best time to put down black polythene is in the spring, and by late autumn all the annuals and many of the perennial weeds, particularly those with shallow roots, such as couch grass and creeping buttercup, will be dead. However, it will take more than one growing season to clear weeds with taproots or very deep roots, while those with corms, such as bulbous buttercup, or bulbils like oxalis, are very persistent.

To *suppress* weeds between shrubs and vegetables, more attractive materials can be used. These include materials which can eventually be dug in, such as compost, composted manure, peat and leaf mould, and materials which have a high carbon content which are best not dug in but left to gradually on the surface. These include straw, pine needles, shredded bark and woodchips.

A layer thick enough to stop weeds from germinating will be needed. The thickness will be determined partly by the material being used and partly by the size of the plant being mulched. Small plants should not be swamped, although large or tall plants will grow happily in a mulch 15cm (6 in.) deep. In most cases 10cm (4 in.) is ideal.

Which mulch to use and where to use it depends on a number of considerations. Compost and composted manures are too valuable for weed control and should only be used around vegetable crops or fruit bushes and roses which require a continuing supply of nutrients. Leaf mould and peat, on the other hand, while providing some nutrients very slowly, do look attractive.

Sawdust, forest bark and shredded woodchips are long-lasting, but should be composted before use and are best used around trees and shrubs.

All mulches should help prevent evaporation of moisture from the soil. Mulches particularly help shallow rooted plants which need a constant supply of moisture. But it is important to wet the soil thoroughly before applying a mulch, since rain-water will only percolate through slowly and may be absorbed before it reaches the soil. Mulches also help to minimize the ill-effects of extreme fluctuations in temperature. They are best applied during the growing season. If they are applied during the winter or early spring when the soil is cold they may prevent the soil warming up and therefore slow down plant growth. In the summer they help prevent the soil from drying out, and if applied in autumn they help keep the soil warm.

Mulches also encourage earthworms which create channels for drainage and aeration as well as working the mulch into the soil.

Crops planted through black horticultural polythene will have few weeding problems. Prepare the ground thoroughly before covering the bed with polythene and securing the edges by burying them in the soil. Slits can then be cut with a pair of scissors or knife through which plants can be planted.

The use of black polythene need not be restricted to vegetables; ornamentals or alpines can also be planted through black polythene and covered with bark or gravel to give an attractive finish.

In conservation areas or rough areas where weeding is impracticable the establishment of trees and hedges can be greatly enhanced by planting through black polythene.

Hedge saplings can be planted through slits in black polythene in a similar manner to vegetables. Trees will need at least a square metre (yard) or black polythene around their base.

Black polythene is impermeable and will conserve water in the soil. However, it will not allow any rain through either, which makes watering more of a problem. Water or liquid feed will need to be applied through the planting holes. Alternatively, an irrigating hose can be laid beneath the polythene from the outset.

To overcome this problem a number of products have been developed which are permeable to air, water and liquid feeds but dense enough to impede weed growth. Such material consists of a finely woven black plastic film or porous fabric. This is generally more expensive than black polythene, but does save on time and labour.

A biodegradable product made from compressed peat and cellulose will keep the weeds down for two or three months, by which time the vegetables should be mature or sufficiently established to compete with the weeds. This is supplied as a roll of stiff brown paper and is used in the same way as black polythene. However, it does have a number of drawbacks: firstly it tends to shrink when first wetted so it should be given a good soaking before planting; secondly, it can be difficult to keep down in windy situations; and thirdly, it tends to decay at the edges where it is buried in the soil. It can be used instead of newspaper under other mulches as described above and is completely free of lead.

Cover crops

Green manuring is a method of improving the fertility of soils by adding organic matter. By incorporation into the soil and subsequent decomposition, nutrients become available to the next crop. Green manures also help improve the soil structure, increase biological activity in the soil and help prevent nutrient leaching if planted as a winter crop.

A green manure crop takes up space and reduces the amount of light reaching the soil, thus preventing weeds establishing themselves. Green manure crops can also be used either as a cleaning crop, in the same way as potatoes are used, or sown between crops to reduce weed problems.

Undersowing crops is desirable as long as the green manure does not significantly compete with the crops in which it is sown. Although undersowing is principally used on small-scale farms and horticultural systems, there is no reason why it cannot be adapted for use in larger vegetable gardens.

Undersowing with green manures is valuable for a number of reasons: they provide cover for predatory beetles, such as rove beetles and ground beetles; they compete with pernicious weeds such as bindweed and horsetail and with late weeds such as fat hen; they increase the nutrient availability for the next crop.

Underplanting with green manures, however, must be related to other crops in the rotation, particularly with respect to pest and disease control. For instance, leguminous green manures should not be followed by peas or beans and mustard should not be followed by brassicas. On the other hand, underplanting brassicas with clovers may reduce cabbage root fly damage and aphid populations (see the relevant intercropping sections).

(See Appendix II for list of green manure crops.)

Herbicides

Ammonium sulphamate is probably the safest weedkiller, as it decays in the ground into sulphate of ammonia. It is safe to plant most crops eight weeks after treatment. It kills couch grass, docks, oxalis and many annual weeds. It will also kill tree stumps. It is non-selective. Spot weeding, especially in lawns, is often effective.

Spot weeding using table salt is also effective.

Flame weeding

Flame weeding is used on small-scale farming enterprises in Europe, particularly in Germany, using quite large and sophisticated equipment. Flame weeding takes off the annual weeds but does not destroy perennial weeds with taproots which tend to re-sprout. A gas burner is particularly useful in controlling weeds especially when they are growing in paths.

Cultural control

Annual weeds must be destroyed before they flower, as disturbing the ripe seed pods will just help spread the seeds. Hoe or hand weed when the soil is dry.

Perennial weeds with creeping stems above ground must be dug out before the creeping stems start to root. Once established they cannot be got rid of by hoeing or hand weeding. Perennial weeds with creeping underground roots or stems can be a real problem, because hand pulling or hoeing is likely to make the problem worse as each little bit of root left in the soil will re-grow. Dig out as much root or stem as possible. Perennials that produce bulbils are also very difficult to control. Each little bulbil left in the ground will re-grow. Pulling them out tends to split up the bulbils and increase the problem. Digging them out is also very difficult, as clumps of bulbils shatter and become dispersed in the soil. Perennials with taproots tend to re-sprout and have very deep roots which are difficult to dig out completely. Spot weedkilling with salt or ammonium sulphamate is probably the only answer, especially in lawns.

6
Plant diseases

Diseases of plants are often caused by microscopic organisms such as fungi, viruses and bacteria; but physiological disorders and nematodes (eelworms) can also cause similar symptoms to appear.

Fungi

Fungi, unlike most plants, lack chlorophyll and cannot manufacture their own food. Instead, they have microscopic threads called hyphae which grow and digest either living or dead matter.

Those which attack dead plant tissue are called saprophytes and are generally useful as they help to recycle organic matter.

Fungi which attack annual or herbaceous perennial plants survive by overwintering in the soil or on dead plants, ready to re-infect growing plants in the spring. Most fungi reproduce and spread by spores. These are produced in enormous numbers in the spring and summer and can be blown considerable distances by the wind or splashed onto leaves by the rain. Spores can live in the soil for long periods, waiting to attack suitable plants.

Cultural control

Prevention is better than cure and much can be done to prevent fungi attacking plants. Sources of infection can be reduced by promptly clearing away any debris from old plants and not allowing prunings, dead leaves and the like to accumulate and by burning diseased plants and not composting them. Disinfect the greenhouse thoroughly once a year by removing all the plants and washing down the frame, glass and staging; and wash and disinfect plant pots, seed boxes, canes and other equipment at the end of the season to reduce infection from fungi.

Do not introduce diseased plants into the garden, especially from the cabbage family which may have club root.

To prevent fungal diseases building up in the soil, vegetables should not be grown year after year in the same ground. Club root and white rot are very persistent and a normal rotation will not help.

Use a mulch of organic material or black polythene over the soil to help avoid the spread of diseases by preventing the spores of fungi splashing up onto plants.

Plants that are growing vigorously are less likely to be vulnerable to disease, so plants should be watered, fed and pruned regularly. Plants put under stress by being planted in the wrong soil and situation, lacking in moisture or over-crowded are most susceptible to disease.

If the same disease problem recurs, planting resistant varieties grown from seed may alleviate the problem. (See Appendix II for a list of resistant varieties.)

Chemical control

Fungicides such as copper salts and dispersible sulphur protect plants by forming a barrier on the plant surface. These fungicides have no effect on deep-seated infections and must be applied accurately and regularly throughout the growing season to be effective.

Copper fungicides are usually presented as mixtures of copper sulphate and slaked lime, called Bordeaux mixture, and copper sulphate and washing soda, called Burgundy mixture.

Copper fungicides appear to be harmless to many parasites and predators. Dispersible sulphur, however, can harm various parasitic wasps and predatory mites, which may lead to increased pest problems. Some people are sulphur sensitive, so protective clothing and masks should be used during application. Sulphur can also damage strawberry plants.

Formaldehyde can be useful as a soil and pot sterilizer as it does not leave harmful residues but, nevertheless, it should be used with care as it is harmful to a number of soil organisms. Worms in particular are affected by formaldehyde.

(See Appendix II for a list of proprietary fungicides.)

Microbial fungicide

Trichoderma viride is a mycoparasite which has an antagonistic effect on certain fungal diseases.

Trichoderma is presented either as a wettable powder or in pellet form. Pellets are applied to the tree during February to April in pre-drilled holes, 5cm (2 in.) deep, which are spaced every 10cm (4 in.) in a spiral around the tree. One hole should penetrate to the centre of the tree where it will attack any rots in the heartwood. Between 2 and 5 pellets, depending on the size of the tree, are placed in each hole.

As a wettable powder, *Trichoderma* can either be mixed into a paste or dissolved to make a spray. In this form it is used to treat tree wounds. On large wounds *Trichoderma* should be applied with a paint brush. On smaller wounds use as a spray. Wounds should always be treated on the same day as the pruning and the mixture must be used within four hours.

Trichoderma is effective against the following organisms:

Tree stem and root rot — *Heterobasidion annosum*
Silver-leaf disease of fruit trees and eucalyptus — *Chondrostereum purpureum*
The internal decay producing organisms of creosoted utility poles — *Lentinus lepideus*
Dutch elm disease — *Ceratocystis ulmi*
Honey fungus — *Armillaria mellea*

Viruses

There are no effective chemical methods of controlling viral induced plant diseases. Where possible use resistant varieties. (See Appendix II.)

Bacteria

A number of rots and cankers are caused by bacteria. The most common and widespread is bacterial soft rot, which affects a number of vegetables in the garden and in store. Turnips, swedes, celery, cucumbers, leeks, lettuces, onions, parsnips, potatoes and tomatoes can all be affected. Ornamentals such as cyclamen, hyacinths and irises are also affected.

Bacterial canker is most serious on cherries, plums and other *Prunus* species, but one type of bacterial canker also affects poplar trees.

Cultural control

Bacterial soft rot can be reduced by increasing the rotation between susceptible crops, avoiding a potassium–nitrogen imbalance and improving the drainage. At the same time pests and diseases which damage plants and allow bacterial soft rot to become established should be controlled.

The only effective method of avoiding canker in poplar trees is to plant resistant varieties; cultivars of black poplar and Lombardy poplar are resistant. Unfortunately, no plum or cherry cultivars are resistant and some, such as the Victoria plum, are very susceptible. Great care must be taken to avoid damage to trees which would allow infection to become established.

Plant diseases and their control

Powdery mildew

Powdery mildew appears as a white, dirty or floury covering on the leaves and shoots of many types of vegetables, fruit and flowers. Each species has its own specific type of powdery mildew. It is especially damaging on apples, gooseberries, currants, peaches, plums, grapevines, turnips and ornamentals such as roses, forget-me-nots, Michaelmas daisies, delphiniums and chrysanthemums.

It often occurs in dry weather, or when plants are suffering from lack of water and when plants are crowded together.

Control

In the winter cut out and burn any infected shoots, cutting back to several buds below visible infection.

During the growing season spray with a sulphur-based fungicide. Sulphur can damage some plants, however.

Downy mildew

This is similar in name only to powdery mildew. The upper surfaces of leaves show yellow or dull patches with a greyish mould growth beneath. It is a problem on lettuces and brassicas but also occurs on spinach and onions. It occurs also on many ornamentals such as antirrhinums, sweet peas, poppies and wallflowers. Downy mildew is a problem of warm, damp weather.

Control
Remove and burn affected leaves and spray with a copper-based fungicide.

Rust
Rust diseases typically form yellow, orange, brown and occasionally black rusty spots on the undersides of leaves and stems. It is typical of hollyhocks and other ornamentals such as antirrhinums, pelargoniums, sweet williams, chrysanthemums, roses and several species of ornamental trees and shrubs. Other plants frequently attacked are plums, leeks, chives and mint.

Control
Remove and burn affected leaves and spray ornamentals with a copper-based fungicide and fruit and vegetables with either a copper- or sulphur-based fungicide.

Leaf spots
Virtually every species of plant is affected by one or more types of fungus which cause spots on the leaves. Leaf spot can present itself as spots, rings or blotches.

The most important spots of ornamentals are those affecting pansies, phlox, polyanthus, poppies, irises, sweet williams, carnations, roses and hellebores, and of vegetables those of broad beans and celery. Currants and gooseberries are also affected by leaf spot.

Control
Remove and burn diseased leaves and spray ornamentals with a copper-based fungicide.

Cankers
Cankers form blister-like lesions on the stems of woody plants. They are caused by bacteria and fungi that grow or multiply very slowly beneath the bark, killing a slightly larger area each year until eventually the entire branch or stem may be girdled — which in the case of the trunk may kill the tree. In gardens cankers are most prominent on apples and pear trees but can also occur on other trees such as poplar and larch and ornamentals such as roses.

Control

Cut out and burn diseased branches and stems. Cut out canker lesions in the winter and paint large surfaces with *Trichoderma* paste. Although this will not kill the canker it will prevent the establishment of other fungal or bacterial diseases.

Scab

Some fungi cause spots on plants which have a crusty appearance and many are given the descriptive name of 'scab'. The commonest and most important are apple and pear scab and potato scab.

Control

Scab on apples and pears can be controlled using Bordeaux mixture. Apple scab can also be controlled using dispersible sulphur.

Potato scab is most common in light alkaline soils which are low in humus. Increasing the acidity of the soil with grass cuttings, leaf mould or manure and watering once the tubers begin to form can help. Some kinds of potato such as Desirée and Maris Piper are particularly susceptible to scab and should be avoided. Perhaps the best way to avoid potato scab is to use resistant varieties. (See Appendix II.)

Smuts

Smuts are characterized by black, sooty spores on the aerial parts of plants. The most important is onion smut and that affecting sweetcorn.

Control

There are no reliable fungicidal treatments for smut. Grow plants in a different part of the garden for at least five (or more) years.

Wilt

Plants affected by wilt suddenly droop and die in hot weather because their water-conducting vessels are blocked. Cucumbers, tomatoes, carnations and chrysanthemums are the plants most likely to be affected.

Control

There are no reliable fungicidal treatments for wilt. Use sterile seed compost and burn badly-affected plants.

Abnormalities

Many diseases caused by fungi or bacteria result in malformed or distorted growth in some part of the plant.

Common examples include club root on the roots of brassicas; crown gall on the stems of trees and shrubs; leaf curl, especially on peaches and almonds; and witches'-broom on birches and other trees.

Peach leaf curl

Peach leaf curl on peach and almond trees can be checked by spraying with a copper fungicide just before the leaves fall in autumn and again before the buds swell in early spring (January or February), and then a fortnight later.

The disease spores overwinter among the bud scales and in the bark of the tree; they germinate in spring and the new spores are spread to the developing leaves by rain-water splash. This is why peaches under glass are rarely affected. Wall-trained trees can be protected by covering them with plastic sheeting during the winter.

Club root

There is no cure for club root, so if your garden is free of club root do not bring in plants of the cabbage family to the garden — raise them from seed.

Club root causes swellings on the roots and the effects can range from a reduction in yields to total crop failure. If the garden has the disease there are some measures that can be taken to improve the crop. Club root does not develop so effectively in a limey soil, so add lime to the plot the season before planting. Always rotate brassicas with a space of at least three years between and raise plants in 10cm (4 in.) pots before planting out so that they have a reservoir of clean soil around the roots. Plant out in a trench filled with compost.

Rots and decays: fungal and bacterial

Fungal diseases often cause the plant's tissues to disintegrate; these diseases are called rots or decays.

Many of these diseases are given individual names — apple

brown rot, potato blight, honey fungus or grey mould for instance.

Blight

Blight affects not only potatoes but also tomatoes. It generally occurs in warm, humid conditions where temperature exceeds 10°C (50°F) and a relative humidity of 75 per cent or more.

Control

There is no cure for potato blight, but Bordeaux mixture will prevent it spreading. If the whole crop becomes infected it is best to remove and burn the foliage to prevent the spores washing into the soil and infecting the tubers.

The tubers may be eaten but should not be used as seed potatoes and you should make sure no potatoes are left in the soil to act as a source of infection for the next crop. Tomato leaf blight can be prevented by dusting the leaves with a copper-based fungicide.

Grey mould (Botrytis)

Botrytis is a fungal disease of many plants and is particularly destructive in wet seasons. It is also a disease of stored crops such as onions.

Botrytis forms a grey fluffy mould on raspberries, strawberries, grapes and currants and is particularly destructive in wet summers. It also attacks the leaves and stems of many bedding plants such as petunias, the blooms of a wide variety of flowers such as chrysanthemums, dahlias and peonies — particularly under humid conditions — and tomatoes and lettuces.

Without using synthetic fungicides, botrytis is very difficult to control. Remove mouldy leaves and flowers immediately and burn badly infected plants.

Onion neck rot (*Botrytis allii*) is very common and widespread and is the commonest disease of stored onions. Infection is seed-borne and is therefore difficult to control unless treated seed is used. Check stored onions for rots and remove damaged ones.

Glasshouse crops should be well ventilated.

Damping off

Damping-off fungi attack the roots and stem bases of seedlings. Rot and shrinkage occurs at ground level and the plants topple over and die.

Sow thinly, never over-water and ensure adequate ventilation under glass. Pots, seed boxes and compost should all be sterilized.

An isolate of *Trichoderma harzianum* has been found to be suitable for the control of damping-off pathogens, but is not commercially available at present.

Lawn diseases

Fusarium or 'snow mould' is the commonest disease of lawns. It shows up as small patches of yellowing grass, mostly in the autumn or in spring after snow has melted. Patches increase in size and may merge, resulting in areas of dead grass. In damp conditions white or pink mould may appear at the edges.

Fusarium can be discouraged by good lawn management and good drainage. It thrives under moist conditions and can be discouraged by reducing shade, avoiding long grass, removing grass clippings and by brushing the lawn to disperse the dew in the morning.

In practice these measures are difficult and hard work. Fusarium can be avoided by not using nitrogen fertilizers excessively, particularly in the autumn, and by keeping the soil on the alkaline side.

Honey fungus

Honey fungus usually enters a garden by colonizing a tree stump. This supplies the fungus with a large reserve of food. It then spreads outwards at a rate of about one metre (yard) each year along the old roots of the tree, until it comes into contact with the roots of a living tree or shrub. It may then attack the roots and kill the plant.

Honey fungus forms characteristic clumps of yellowish toadstools on or around the old stump. A succession of dying shrubs or trees near the stump indicate the presence of honey fungus.

If honey fungus is suspected, a trench 45cm (18 in.) deep should be dug around the old stump to separate the healthy tree from the infected area. The stump should then be burnt away to kill the roots and prevent further infection. Back fill the trench afterwards with clean soil.

After a year or so, re-plant resistant species in the infected area. (See Appendix II for lists of plants susceptible and resistant to honey fungus.)

If it is necessary to retain the infected tree, it is possible to treat it using the microbial fungicide *Trichoderma* (see page 101).

Appendix I
Plants that attract beneficial birds and insects

Plants attractive to butterflies

Spring flowering plants
Alyssum saxatile
Aubrietia
Myosotis (forget-me-not)

Summer flowering plants
Buddleia × *weyeriana*
Buddleia alternifolia
Calendula officinalis (marigold)
Chrysanthemum
Hebe armstrongii
Hebe brachysiphon 'White Gem'
Heliotropium × *hybridum*
Helichrysum bracteatum
Iberis umbellata (candytuft)
Lobelia
Mignonette
Phuopsis stylosa
Red Valerian
Scabiosa atropurpurea
Sweet william
Syringa microphylla
Thyme

Autumn flowering plants
Aster spp. (Michaelmas daisy)
Buddleia davidii
Ceratostigma willmottianum
Chrysanthemum
Escallonia 'Macrantha'

Helenium
Hyssopus officinalis
Lavandula (lavender)
Mentha rotundifolia (apple mint)
Origanum vulgare (wild marjoram)
Sedum spectabile

Wild flowers which are attractive to bees and butterflies as sources of nectar and/or pollen

Birds-foot trefoil (*Lotus corniculatus*)
Bluebell (*Hyacinthoides non-scripta*)
Broom (*Cytisus scoparius subsp. scoparius*)
Lesser celandine (*Ranunculus ficaria*)
Chicory (*Cichorium intybus*)
Cornflower (*Centaurea cyanus*)
Oxeye daisy (*Leucanthemum vulgare*)
Feverfew (*Tanacetum parthenium*)
Honeysuckle or wild woodbine (*Lonicera peridimenum*)
Hound's-tongue (*Cynoglossum officinale*)
Knapweed (*Centaurea nigra*)
Greater knapweed (*Centaurea scabiosa*)
Lady's-smock (*Cardamine pratensia*)
Common mallow (*Malva sylvestris*)
Musk mallow (*Malva moschata*)
Wild marjoram (*Origanum vulgare*)
Marsh-mallow (*Althaea officinalis*)
Heartsease or wild pansy (*Viola tricolor*)
Purple-loosestrife (*Lythrum salicaria*)
Ragged-robin (*Lychnis flos-cuculi*)
Common rock-rose (*Helianthemum nummularium*)
St John's-wort (*Hypericum perforatum*)
Devil's-bit scabious (*Succisa pratensis*)
Field scabious (*Knautia arvensis*)
Soapwort (*Saponaria officinalis*)
Teasel (*Dipsacus fullonum subsp. sylvestris*)
Thrift or sea pink (*Armeria maritima subsp. maritima*)
Wild thyme (*Thymus praecox subsp. ariticus*)
Common toadflax (*Linaria vulgaris*)
Common valerian (*Valeriana officinalis*)
Vervain (*Verbena officinalis*)

Wild wallflower (*Cheiranthus cheiri*)
Yarrow (*Achillea millefolium*)

Trees which attract honeybees

Prunus dulcis (almond)
Malus spp. (crab apple)
Prunus cerasifera (cherry plum)
Cotoneaster frigidus
Robinia pseudo-acacia sempervirens
Corylus spp. (hazel)
Ilex spp. (holly)
Ilex glauca
Aesculus hippocastanum (horse chestnut)
Aesculus carnea
Aesculus indica
Catalpa bignoides (Indian bean tree)
Cercis siliquastrum (Judas tree)
Tilia × *europaea* (common lime)
Tilia platyphyllos
Tilia cordata

Tilia × *euchlora* *Tilia insularis* *Tilia maximowicziana* *Tilia miquellana* *Tilia mongolica* *Tilia tomentosa*	Limes not attacked by aphids or which produce suckers

Acer saccharum (sugar maple)
Mesphilus germanica
Pyrus spp. (ornamental pears)
Ligustrum sinensis
Ligustrum vulgare
Ligustrum ovalifolium (common privet)
Castanea sativa (sweet chestnut)
Ailanthus altissima (tree of heaven)
Salix caprea (goat willow)
Salix alba (white willow)

Beneficial birds and their food

Bird	**Food**
Barn owl	small rodents

Blackbird	invertebrates, seeds, berries, fruit
Blackcap	insects, berries, fruit
Blue tit	insects, seeds
Chaffinch	invertebrates, fruit
Coal tit	invertebrates, seeds, nuts
Dunnock	insects, seeds
Fieldfare	invertebrates, berries
Goldfinch	seeds (thistles etc.)
Great tit	insects, seeds, nuts
Green woodpecker	insect larvae (trees)
House martin	insects
Jackdaw	plant and animal material
Little owl	small rodents, insects
Long-tailed tit	insects, seeds
Magpie	insects, small rodents
Mistle thrush	fruit, invertebrates, snails
Pied flycatcher	insects
Pied wagtail	insects
Redstart	insects
Redwing	invertebrates
Robin	invertebrates, seeds, berries
Song thrush	invertebrates, berries, seeds
Spotted flycatcher	insects (flying)
Starlings	invertebrates, seeds, berries, fruit
Swallow	insects (flying)
Swift	insects (flying)
Tree creeper	insects (tree)
Wren	invertebrates
Yellowhammer	insects, seeds, fruit

Food plants for birds

This list is not exhaustive — the RSPB have formulated a wild flower mixture which is available upon request.

Berberis
Blackberry
Elderberry
Guelder Rose (*Viburnum opulus*)
Hawthorn
Holly
Honeysuckle
Michaelmas Daisy

Rowan
Scabious
Sea Buckthorn
Snowberry
Spindle
Sunflower
Wayfaring tree (*Viburnum lantana*)
Yew

Appendix II
Disease resistant varieties of vegetables and fruit

(Check seed catalogues for availability.)

Beetroot
Bolt resistant
Boltardy, Monopoly, Detroit-Rubidus, Regala, Bikores, Beethoven, Slowbolt
Good storage performance
Boltardy, Early Wonder

Runner beans
Bean mosaic virus and Anthracnose resistant
Red Knight

Dwarf French beans
Good disease resistance
Aramis

Brussels sprouts
Powdery mildew resistant
Cor, Porter, Rampart, Achilles, Camilla, Aries, Sheriff, Glentora, Gabion
Ringspot resistant
Adeline, Cor, Gabion, Lunet
White blister resistant
Fortress, Mallard

Cauliflowers
Good tolerance to downy mildew
Autumn Giant-Late Supreme, Snow February, Snow January, Surf Rider
Bolt resistant
Alpha Paloma

Cold tolerant and not prone to blindness
Mill Reef

Cabbages
Good resistance to turnip mosaic virus
Multiton hybrid, Sphinx hybrid, Zerlina hybrid

Celery
Bolt resistant
Lathom Self-blanching

Onions
Bolt resistant
Sturon
Good storage performance
Rijnsburger, Southport, Yellow Globe, Hygro, Caribo

Parsnips
Canker resistant
Avonresister, Gladiator, Marrow Improved, Alba, White Gem

Carrots
Good storage performance
Chantenay Red Cored, Autumn King Vita Longa
Partial resistance to carrot fly
Sytan, Tip-top, Early Scarlet Horn, French Forcing Horn, James Intermediate, Bridge, Empire, Fancy, Ideal, Prima, Tantal, Touchon, Vertou
Good cavity spot resistance
Nantura

Swedes
Good tolerance to club root and mildew
Marian

Turnips
Good storage performance
Tokyo Cross

Spinach
Mosaic virus resistant
Melody

Asparagus
Rust resistant
Mary Washington

Potatoes

Eelworm resistant
(mainly the golden species of potato cyst nematode)
Cara, Maris Piper, Pentland Javelin, Pentland Meteor, Costella, Kingston, Sante, Aminca

Slug resistant
Stourmont Enterprise, Pentland Falcon, Pentland Dell

Gangrene resistant
Home Guard, Wilja, Maris Peer, Kingston, Maris Piper

Foliage blight resistant
Ulster Sceptre, Estima, Wilja, Maris Peer, Cara, Romano, Record

Common scab tolerant
Arran Comet, Pentland Javelin, Maris Peer, Kingston, Pentland Crown

Leaf roll virus resistant
Aminca, Pentland Crown, Sante

Virus Y resistant
Maris Bard, Pentland Javelin, Cara, Desirée, Pentland Crown, Romano, Sante

Spraing-tobacco rattle virus resistant
Aminca, Home Guard, Romano, Record

Tuber blight resistant
Ulster Sceptre, Wilja, Cara, Record, Sante

Black leg resistant
Ailsa

Lettuce

Downy mildew resistant and lettuce mosaic virus tolerant
Capitan, Cindy, Clarion, Mirena, Nancy, Oresto

Root aphid tolerant
Sabine, Avoncrisp, Avondefiance, Lakeland

Lettuce mosaic virus tolerant
Calona, Malilsa

Tip burn resistant
Cindy, Clarion, Mirena, Patly, Soraya, Calona, Novita

Cucumbers

Powdery mildew resistant
Amslic

Tomatoes

Cladosporum resistant
Odine F1 hybrid, Ronaclave

Verticillium resistant
Ronaclave
Resistant to cracking
Carmelo
Resistant to root-rot
Celebrity, Piranto
Tobacco mosaic virus resistant
Shirley

Raspberries
Partial resistance to aphids
Malling Delight, Malling Leo, Malling Orio

Cape gooseberries
Good storage performance
Golden Berry

Honey fungus

Susceptible plants
Apple
Birch
Cedar
Cherry
Cypress
Lilac
Monkey puzzle
Pear
Pine
Plum
Privet
Rhododendron
Viburnum
Walnut
Wellingtonia
Willow

Resistant plants
Bamboo
Beech
Box
Clematis
Cotinus (smoke tree)
Douglas fir
Elaeagnus
Elder
Holly
Honeysuckle
Ivy
Larch
Lime
Mahonia
Rhustiplina
Robinia
Rock rose
Silver fir
Tamarisk
Thorn
Yew

Green manure crops

Alfalfa or Lucerne (*Medicago sativa*) — nitrogen fixer
Buckwheat (*Fagopyrum esculentum*)
Chicory (*Cichorium intybus*)
Red clover (*Trifolium pratense*) — nitrogen fixer
White clover (*Trifolium repens*) — nitrogen fixer
Lupin, annual blue (*Lupinus augustifolius*) — nitrogen fixer; fungi in roots also make phosphorus available
Lupin, annual white (*Lupinus albus*) — nitrogen fixer; fungi in roots also make phosphorus available
White mustard (*Sinapsis alba*)
Pea, forage or field (*Pisum arvense*) — nitrogen fixer
Radish, fodder (*Raphanus sativus*)
Grazing rye (*Secale cereale*)
Winter tares or common vetch (*Vicia sativa*)

Proprietary insecticides, fungicides and herbicides

Many of these products or similar products are available at ordinary gardening centres or from organic gardening suppliers by post (see Appendix III for list of addresses). Product names change but their contents remain the same — check the pack label of any product for the right insecticide or fungicide.

Insecticides

Aerosols

Pyrethrum	Boots' Garden Insect Killer Py Garden Insecticide (Synchemicals)

Dusts

Derris	Abol Derris Dust (ICI) Corry's Derris Dust (Synchemicals) Doff Derris Dust Murphy Derris Dust Tesco Derris Dust Derris Dust (Battle Hayward and Bower)

Pyrethrum	Anti-Ant Duster (pbi) Py Powder (Synchemicals)
Liquid Sprays	
Derris	Liquid Derris (pbi)
Derris/quassia	Bio Back to Nature Insect Spray
Nicotine	Bentley's Nicotine Insecticide
Pyrethrum	Py Spray Garden Insect Killer (Synchemicals) Py Spray Whitefly Killer (Synchemicals) Bio Sprayday
Baits	
Borax	Nippon Ant Destroyer Panant (pbi)
Insecticidal soaps	Savona (Koppert) Safer's Insecticidal Soap Sebon (Biological Control Systems Ltd)
Microbial Insecticides	Vertalec — microbial aphicide (Bunting & Sons) Mycotal — whitefly control (Bunting & Sons) Bactospeine — *Bacillus thuringiensis* (Koppert) Biobit — *Bacillus thuringiensis* (Bunting & Sons) *Heterorhabditis* (nematode against weevil) (Koppert)

Combined insecticides and fungicides

Derris/sulphur	Bio Back to Nature Pest and Disease Duster

Fungicides

Liquid sprays	Bordeaux Mixture (Battle Hayward and Bower)
Copper compounds	Corry's Bordeaux Mixture (Synchemicals) Murphy Liquid Copper Fungicide
Sulphur	Corry's Green Sulphur Corry's Yellow Sulphur

	Koppert's Natural Garden Fungicide
	Sulphur W80 (Intercrop Ltd)

Herbicides

Ammonium sulphamate	Amicide

	Koppert's Natural Garden Fungicide
	Sulphur W80 (Intercrop Ltd)

Herbicides

Ammonium sulphamate	Amicide

Appendix III
Useful addresses

Biological control
Biological Control Systems Ltd
Treforest Industrial Estate
Pontypridd
Mid Glamorgan CF37 5SU
Tel: 0443 841155
(For pheromone traps)

Bunting & Sons
The Nurseries
Great Horkesley
Colchester
Essex CM6 4AJ
Tel: 0206 271300

English Woodland Ltd
Hoyle Depot
Graffham
Petworth
West Sussex GU28 0LR
Tel: 07986 574

Koppert (UK) Ltd
Biological Control
PO Box 43
Tunbridge Wells
Kent TN2 5BX

National Centre for Organic Gardening
Ryton-on-Dunsmore
Coventry CV8 3LG
Tel: 0203 303517
(For Agryl P17)

Natural Pest Control
'Watermead'
Yapton Road
Barnham
Bognor Regis
West Sussex PO22 0BQ
Tel: 0243 553250

Butyl liners
Butyl Products Ltd
Radford Crescent
Billericay
Essex CM12 0DW
Tel: 0277 653281

Stapeley Water Gardens Ltd
Stapeley
Nantwich
Cheshire CW5 7LH
Tel: 0270 623868

Earwig nests
Langdon (London) Ltd
5 Worminghall Road
Ickford
Aylesbury
Bucks HP18 9JJ

Insecticide manufacturers and suppliers
(Most of these products can be supplied by organic gardening suppliers.)

BT4000 (Bacillus thuringiensis)
Clavering Organics Real Gardening Ltd
5 Chaucer Industrial Estate
Dittons Road
Polegate
East Sussex BN26 6JF

Binab
Binab
Box 54
S-19300 Sigtuna
Sweden

Hortopaper
Donaldson Paper & Board Sales Ltd
Suite 9
Essex House
15 Station Road
Upminster
Essex RM14 2SJ
Tel: 0422 51300

Liquid derris insecticides (greenfly spray)
Soft soap
Battle Hayward & Bower Ltd
Victoria Chemical Works
Lincoln
Tel: 0522 29206

Nicotine insecticide
(XL-All-Insecticide, Synchemicals)
(Available to professional growers only)
Joseph Bentley Ltd
Barrow on Humber
South Humberside
Tel: 0469 31565

Non-drying sticky adhesive — Trappit
Agralan
The Old Brickyard
Ashton Keynes
Swindon
Wiltshire SN6 6QR
Tel: 0285 860015

Papronet
Propropac Sales Ltd
Wyke Works
Hedon Road
Hull HU9 5NL

Pure pyrethrum powder
Chase Organics (GB) Ltd
Addlestone
Weybridge
Surrey KT15 1HY
Tel: 0932 58511

Quassia chips
Brome & Schimmer Ltd
Greatbridge Road (Industrial Estate)
Romsey
Hants SO5 0HR
Tel: 0794 515595

Insecticidal Soaps
Safer Agro-Chem Ltd
455 Milner Avenue
Unit 1
Scarborough
Ontario
Canada M1B 2K4

Koppert (UK) Ltd
P.O. Box 43
Tunbridge Wells
Kent
TN2 5BX

Organic gardening suppliers

Chase Organics (GB) Ltd
Addlestone
Weybridge
Surrey KT15 1HY
Tel: 0932 58511

Cumulus Organics Ltd
Timber Road
Two Mile Lane
Highnam
Gloucester GL2 8BR
Tel: 0452 305814

Intracrop Ltd
The Crop Centre
Waterstock
Oxford OX9 1LJ
Tel: 08447 377
(For fungicides etc.)

National Centre for Organic Gardening
Ryton-on-Dunsmore
Coventry CV8 3LG
Tel: 0203 303517

Wildflower seeds etc.

John Chambers
15 Westleigh Road
Barton Seagrave
Kettering
Northants NN15 5AJ
Tel: 0933 681632
(Wildflower seeds, bee plants, green manure plants, etc.)

Kingsfield Tree Nursery
Broadenham Lane
Winsham
Chard
Somerset
Tel: 046030 697
(Native trees, shrubs and wild flowers)

Yellow sticky pads (Aeroxon)
There is no UK manufacturer of these, but they can be supplied by the National Centre for Organic Gardening (address given above) and many gardening centres or from the Natural Gardening Research Centre (address below)

Yellow sticky traps
Natural Gardening Research Centre
State Highway No 46
P.O. Box 300
Sunman, IN 47041
USA

Useful Organizations

The Soil Association
86 Colston Street
Bristol
BS1 5BB

Organic Growers Association
86 Colston Street
Bristol
BS1 5BB

Elm Farm Research Centre
Hamstead Marshall
Nr. Newbury
Berkshire
RG15 0HR

NIAB
Huntingdon Road
Cambridge
CB3 0LE
(Impartial advice on vegetable varieties)

Institute of Horticultural Research
(Formally Glasshouse Crops Research Institute)
Worthing Road
Little Hampton
West Sussex
BN17 6LP

IFOAM
Rodale Research Centre
Box 323, R.D.I.
Kutztown, P.A. 19530
USA

Nature et Progrès
91730 Chamarande
France

The Bio-Integral Resource Centre (BIRC)
P.O. Box 7414
Berkeley
CA 94707
USA

Organic Growers Association, W.A.
P.O. Box 213
Wembley
W.A. 6014
Australia

Select bibliography

Baines, C. (1985) *How to Make a Wildlife Garden*, Elm Tree Books

Briggs, J.B. (1965) Biology of some ground beetles (Col.: Curabidae) injurious to strawberries, *Bulletin of Entomological Research*, 56: 79–93

Buczacki, S. (1985) *Beat Garden Pests and Diseases*, Penguin Books

Chinery, M. (1986) *Garden Creepy Crawlies*, Whittet Books

Cromartie, W.J. (1975) The effect of stand size and vegetational background on the colonization of cruciferous plants by herbivorous insects, *Journal of Applied Ecology*, 12: 517–32

Dempster, J.P. (1969) Some effects of weed control on the numbers of the small cabbage white butterfly (*Pieris rapae* L.) on Brussels sprouts, *Journal of Applied Ecology*, 6: 339–45

Feeny, P. *et al.* (1970) Flea beetles and mustard oils: host plants specificity of *Phyllotreta cruciferae* and *Phyllotreta striolata* adults (Coleoptera: Chrysomelidal), *Annals of the Entomological Society of America*, 63(3): 832–41

Gardening from Which? (April 1984) Insect pests: how to keep them under control

Gardening from Which? (April 1985) Plant diseases

Gommers, F.J. (1973) Nematicidal principles in compositae, Veenman and Zonen, Wageningen

Hamilton, G. (1987) *Successful Organic Gardening*, Dorling Kindersley

Harrowsmith (ed.) (1984) Nematode nemesis, *Harrowsmith*, 8 (53): 125–6

Hostettman, K. *et al.* (1982) Molluscicidal properties of various saponins, *Plant Research*, 44: 34–5

Naturalist's Handbook Series, Richmond Publishing Company Ltd, Slough:

No. 3 — *Solitary Wasps* by Peter F. Yeo and Sarah A. Corbet
No. 5 — *Hoverflies* by Francis S. Gilbert
No. 6 — *Bumblebees* by Oliver E. Prys-Jones and Sarah A. Corbet
No. 7 — *Dragonflies* by Peter L. Miller
No. 8 — *Common Ground Beetles* by Trevor G. Forsythe
No. 10 — *Ladybirds* by Michael Majerus and Peter Kearns
No. 11 — *Aphid Predators* by Graham E. Rotheray

N.V.R.S. (1984 No. 10 in N.V.R.S. Series B of *Science in the Vegetable Garden*:
Series B: 1 — *Cabbage root fly in the vegetable garden*
Series B: 9 — *Pests of brassica crops*
(Other titles available in both Series A and B.)

O'Donnell, M.S. and Coaker, T.H. (1975) Potential of intercrop diversity for the control of brassica pests. Proceedings of the Eighth British Insecticide and Fungicide Conference, pp.101–5

Ryan, J. *et al.* (1980) The effect of interrow plant cover on populations of the cabbage root fly *Delia brassicae* (Wiedemann) *Journal of Applied Ecology*, 17: 31–40

Schaufelberger, D. and Hostettman, K. (1983) On the molluscicidal activity of tannin containing plants *Plant Research*, 48: 105–7

Smith, J.G. (1976) Influence of crop background on aphid and other phytophagous insects on Brussels sprouts *Annals of Applied Biology*, 83: 1–13

Symondson, W.O.C. (1989) Biological control of slugs by carabids. 1989 B.C.P.C. Mono No. 41: 295–300

Tukahirua, E.M. and Coaker, T.H. (1982) Effect of mixed cropping on some insect pests of brassicae infestations and influences on epigeal predators and the disturbance of oviposition behaviour in *Delia brassicae. Entomol. Exper. Applic.*, 32: 129–40

Uvah, I.I.I. and Coaker, T.H. (1984) Effect of mixed cropping on some insect pests of carrots and onions *Entomol. Exper. Applic.*, 36: 159–67

Index